INTEGRATED PEST MANAGEMENT FOR

COTTON

IN THE WESTERN REGION OF THE UNITED STATES

WESTERN REGIONAL INTEGRATED PEST MANAGEMENT PROJECT
UNIVERSITY OF CALIFORNIA STATEWIDE INTEGRATED PEST MANAGEMENT PROJECT
UNIVERSITY OF ARIZONA STATEWIDE INTEGRATED PEST MANAGEMENT PROJECT
NEW MEXICO STATE UNIVERSITY

UNIVERSITY OF CALIFORNIA, DIVISION OF AGRICULTURE AND NATURAL RESOURCES
PUBLICATION 3305 $15.00

PRECAUTIONS FOR USING PESTICIDES

Pesticides are poisonous and must be used with caution. READ THE LABEL BEFORE OPENING A PESTICIDE CONTAINER. Follow the label precautions and directions, including requirements for protective equipment. Use a pesticide only against pests specified on the label or in published University recommendations. Apply pesticides at the rates specified on the label or at lower rates if suggested in this publication. Pesticide laws and regulations change frequently; be sure the publication you are using is up to date.

Legal Responsibility. The user is legally responsbile for any damage due to misuse of pesticides, including possible effects of drift, runoff, or residues.

Transportation. Do not ship or carry pesticides with food or feed in a way that allows contamination. Never transport pesticides in a closed passenger vehicle or cab.

Storage. Keep pesticides in original containers until used. Store them in a locked cabinet, building, or fenced area where they are not accessible to children, unauthorized persons, pets, or livestock. DO NOT store pesticides with food, feed, fertilizers, or other materials that may become contaminated by the pesticides.

Container Disposal. Dispose of empty containers carefully; never reuse them. Keep containers away from children and animals. Never dispose of containers where they may contaminate water supplies or natural waterways. Consult local agricultural authorities for procedures for disposing of large quantities of empty containers.

Protection of Non-Pest Animals and Plants. Many pesticides are toxic to desirable animals, including honeybees, natural enemies, fish, domestic animals, and birds. Crops and other plants may also be damaged by mis-applied pesticides. Take precautions to protect non-pest species from direct exposure to pesticides and from contamination due to drift, runoff, or residues. Certain rodenticides may pose a special hazard to animals that eat poisoned rodents.

Posting Treated Fields. For some materials, re-entry intervals are established to protect workers. Keep workers out of the field for the required time after application and, when required by regulations, post the treated area with signs indicating the safe re-entry date.

Harvest Intervals. Some materials or rates cannot be used in certain crops within a specified time before harvest. Follow pesticide label instructions and allow the required time between application and harvest.

Permit Requirement. Many pesticides require a permit from the county agricultural commissioner or other local authority before possession or use. When such materials are recommended in this publication, they are marked with an asterisk (*).

Processed Crops. Some processors will not accept a crop treated with certain chemicals. If your crop is going to a processor, be sure to check with the processor before applying a pesticide.

Crop Injury. Certain chemicals may cause injury to crops (phytotoxicity) under certain conditions. Always consult the label for limitations. Before applying any pesticide, take into account the stage of plant development, the soil type and condition, the temperature, moisture, and wind. Injury may also result from the use of incompatible materials.

Personal Safety. Follow label directions carefully. Avoid splashing, spilling, leaks, spray drift, and contamination of clothing. NEVER eat, smoke, drink, or chew while using pesticides. Provide for emergency medical care IN ADVANCE as required by regulation.

ORDERING
Single copies of this publication, #3305, may be ordered by sending check or money order for $15.00 payable to The Regents of the University of California to:

Publications
Division of Agriculture and Natural Resources
University of California
6701 San Pablo Avenue
Oakland, CA 94608-1239
(415) 642-2431

Information on foreign orders and quantity discounts may be obtained by contacting the above office.

Contributors and Acknowledgments

This manual was produced under the auspices of the University of California Statewide IPM Project with assistance and additional funding provided by the University of Arizona Statewide IPM Project and Regional Research Project W-161, "Integrated Pest Management for Semiarid, Dryland, and Irrigated Agroecosystems in the Western Region."

Prepared by the IPM Manual Group, Mary Louise Flint, Director, at the University of California, Davis. The IPM Manual Group is an office of the Statewide IPM Project.

Paul A. Rude, Senior Writer
Jack Kelly Clark, Principal Photographer

Technical Coordinators

Joe Ellington, Professor of Entomology, New Mexico State University, Las Cruces

Alan G. George, Cotton Farm Advisor, University of California Cooperative Extension, Visalia

Harold M. Kempen, Weed Management Farm Advisor, University of California Cooperative Extension, Bakersfield

Thomas A. Kerby, Cotton Specialist, University of California Cooperative Extension, Shafter

Leon Moore, Coordinator, University of Arizona Statewide IPM Project, Tucson

B. Brooks Taylor, Cotton Specialist, University of Arizona Cooperative Extension, Tucson

L. T. Wilson, Associate Professor of Entomology, University of California, Davis

Contributors

Abbreviations following contributors' names stand for these institutions: (ARS) –United States Department of Agriculture, Agricultural Research Service; (UC)—University of California; (NM)—New Mexico State University; (UA)—University of Arizona.

Agronomy: Robert E. Briggs (UA), Charles R. Farr (UA), Carl V. Feaster (ARS), James L. Fowler (NM), Alan G. George (UC), Donald W. Grimes (UC), Thomas A. Kerby (UC), Jack R. Mauney (ARS), B. Brooks Taylor (UA), Bill L. Weir (UC), and the late O. D. McCutcheon (UC)

Entomology: Louis A. Bariola (ARS), J. Hodge Black (UC), Vernon E. Burton (UC), George D. Butler, Jr. (ARS), Joe Ellington (NM), Daniel Gonzalez (UC), Thomas J. Henneberry (ARS), Roger T. Huber (UA), Thomas F. Leigh (UC), Peter D. Lingren (ARS), Thomas A. Miller (UC), Leon Moore (UA), Harold T. Reynolds (UC), Vahram Sevacherian (UC), Vernon M. Stern (UC), Nick C. Toscano (UC), Theo F. Watson (UA), L. T. Wilson (UC)

Nematology: Howard Ferris (UC), Edward L. Nigh, Jr. (UA), Philip A. Roberts (UC)

Plant Pathology: Lee J. Ashworth, Jr. (UC), Larry F. Benoit (UC), James E. DeVay (UC), Richard H. Garber (ARS), Richard B. Hine (UA), Edward N. Mulrean (UA), Thomas E. Russell (UA), William C. Schnathorst (ARS,UC), Emroy L. Shannon (NM)

Vertebrate Management: James E. Knight (NM), Rex E. Marsh (UC), John L. Stair (UA)

Weed Science: David W. Cudney (UC), Bill B. Fischer (UC), E. Stanley Heathman (UA), Paul E. Keeley (ARS), Harold M. Kempen (UC), Ronald N. Vargas (UC)

Special Thanks

The IPM Manual Group extends special thanks to James M. Lyons, Director of the California Statewide IPM Project; Howard Ferris, Associate Director; Leon Moore, Coordinator of the University of Arizona Statewide IPM Project; and Gary A. McIntyre, Coordinator of the Western Regional 161 Project. Thanks also to Agriculture and Natural Resources Publications, University of California, for the assistance of the editor and artists.

The following persons have generously provided information, offered suggestions, reviewed draft manuscripts, or assisted in obtaining photographs:
J. R. Abernathy, G. Andersen, W. P. Anderson, E. L. Atkins, G. Ballmer, J. L. Baritelle, R. J. Barker, W. W. Barnett, R. Bassett, D. L. Bath, C. A. Beasley, A. Bell, M. R. Bell, W. J. Bentley, D. Bergman, L. L. Bozeman, R. F. Brewer, R. Bugg, D. N. Byrne, J. R. Carey, R. Coviello, T. Dennehy, J. E. Duffus, T. D. Eichlin, C. L. Elmore, L. M. English, D. C. Erwin, L. A. Falcon, C. Finnell, H. M. Flint, K. E. Fry, J. M. Gillespie, C. R. Glover, P. B. Goodell, W. L. Gould, H. M. Graham, J. Granett, G. Guinn, N. Gunther, K. S. Hagen, K. D. Hake, K. C. Hamilton, J. K. Haworth, R. Hoover, O. C. Huisman, A. H. Hyer, E. C. Jorgenson, D. G. Kelly, D. L. Kittock, B. Kobbe, F. F. Laemmlen, M. Lame, W. H. Lange, A. Las, D. L. Lindsey, G. M. Loper, V. Maggi, B. J. McCaskill, M. A. McClure, M. V. McKenry, P. S. McNally, D. S. Mikkelsen, J. Miller, L. Moug, E. T. Natwick, M. R. Nelson, A. O. Paulus, D. A. Pennington, B. Peterson, P. J. Pinter, Jr., R. W. Rosander, E. Seale, N. T. Sego, W. Shelton, I. J. Shields, T. Sienno, R. T. Staten, M. W. Stimmann, S. H. Thomas, I. J. Thomason, P. Trichilo, D. Tuttle, L. A. Urie, R. Van Steenwyk, A. Waiss, D. Weaver, J. W. Whitworth, F. D. Wilson, P. F. Wynholds, F. G. Zalom, M. L. Zavala, L. Zelinski

Information in this manual is based on research sponsored by the University of California, the University of Arizona, New Mexico State University, the U.S. Department of Agriculture Cooperative Extension IPM Fund, and the cotton industry.

Production

Manuscript Preparation: Betty Rudd
Editing: Heidi Seney
Drawings: Marvin Ehrlich and Pamela Fabry
Design and Production Coordination: Naomi Schiff

Contents

Integrated Pest Management for Cotton

This manual is designed to help growers, pest control advisors, and farm managers apply the principles of integrated pest management (IPM) to cotton crops in the Western Region of the United States: California, Arizona, and New Mexico. IPM focuses on the crop plant as a system for channeling the sun's energy into harvestable yield. In an IPM program, pest management is coordinated with production practices to achieve economical protection from pest injury while minimizing hazards to crops, human health, and the environment. The emphasis is on anticipating and preventing problems whenever possible.

The first chapter of this manual, a summary of crop growth and development, provides background for understanding how pest injury interacts with other factors to affect crop production. The outline of cultural practices that follows places integrated pest management in perspective with other practices. Sections on specific pests stress biological information that relates to management strategy and the pest's impact on the crop. The photos and descriptions will help you recognize pests, diseases, and natural enemies. Charts and forms are presented to help you organize pest management activities. The Glossary includes technical terms and other expressions that may be unfamiliar to some readers.

Pesticides are discussed where appropriate, but because pesticide registrations often change, it is impossible to make specific recommendations. However, the References page at the end of the manual includes current recommendation pamphlets from Cooperative Extension Services in the three western states. Stay up to date on pest management methods by keeping in touch with your farm advisor or county agent. Research is continuing in all fields of cotton pest management; monitoring and control methods may change as new information becomes available and as the cotton industry evolves to accommodate changes in technology and market conditions.

The Western Region (Figure 1) produces more than one third of the nation's crop on about 17% of the total acreage. Two species of cotton are grown in the region. The vast majority of the acreage is planted to upland cot-

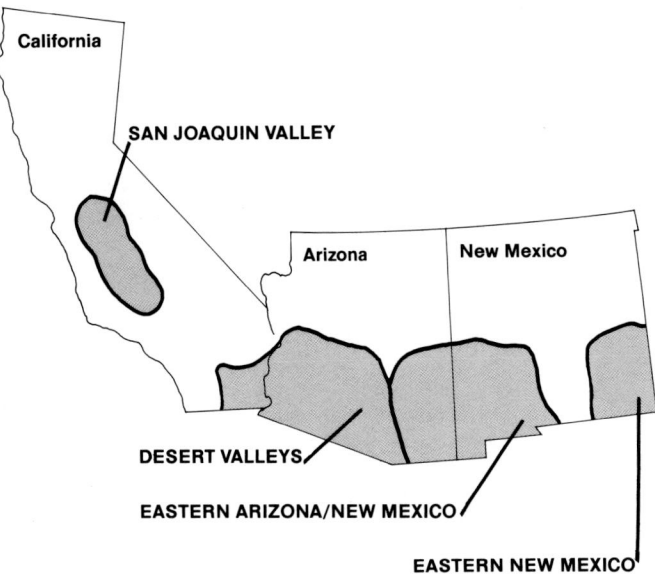

Figure 1. Major cotton production areas in the Western Region.

ton, *Gossypium hirsutum*, but significant amounts of American Pima cotton, *G. barbadense*, are grown in Arizona and New Mexico. The two species are often discussed in terms of the length of the staple or fiber produced. Upland cotton, known worldwide as long staple cotton, is generally called *short staple* in the Western Region. Pima cotton, known worldwide as extra long staple, is called *long staple* in the Western Region.

Except for a small acreage in eastern New Mexico, all western cotton is irrigated. Generally planted in single rows on beds 38 to 40 inches apart, nearly all of it is harvested with spindle pickers. The growing season ranges from 160 days in parts of New Mexico to more than 300 days in the desert valleys of Arizona and southern California; the length of the season is a major factor affecting cultural practices and pest management.

In California's San Joaquin Valley, the largest production area in the Western Region, up to 1.5 million acres of cotton are planted annually. Average yield is about 1000 pounds of lint per acre. To maintain high lint quality and a premium price, cotton culture is limited by law to approved Acala varieties of upland cotton. Major pest problems are spider mites, lygus bugs, seedling diseases, Verticillium wilt, and, as in all areas, weeds. Root-knot nematodes are important in the sandier soils, especially in the Valley's southern end. A disease complex involving root-knot nematodes and Fusarium wilt has been very damaging in some fields.

The desert valleys of southern California and western Arizona have the nation's longest growing season. At lower elevations, as in the Imperial and Yuma valleys, cotton is planted as early as mid-February and harvested from October through January. Upland cotton occupies up to 600,000 acres in Arizona, mostly in the desert valleys, and up to 100,000 acres in southern California. Deltapines are the major varieties. Average yield in Arizona is about 1100 pounds of lint per acre, but in some low desert fields yields may exceed 2000 pounds. Arizona is the leading producer of Pima cotton; up to 42,000 acres are planted and the average yield is about 760 pounds per acre. In the deserts, late-season insect pests, especially pink bollworm and tobacco budworm, are major problems. Important diseases include boll rots, cotton root rot, and seedling diseases.

Up to 125,000 acres of cotton are planted in the area that includes eastern Arizona and western and central New Mexico. Growing conditions are similar to those in the northern San Joaquin Valley; the season is much shorter than in central Arizona. Acala varieties of upland cotton predominate, but there is also some Pima cotton. Major pest problems are lygus bugs, seedling diseases, Verticillium wilt and, in sandy soils, root-knot nematodes. Pink bollworm occurs regularly in eastern Arizona, but it has fewer generations and is less damaging there than in the desert valleys.

Conditions in eastern New Mexico are more similar to those in neighboring parts of Texas than to the rest of the Western Region. Much of the cotton acreage is planted to Acala varieties, but an increasing proportion is planted to varieties suitable for stripper rather than spindle harvesting. Some cotton in this area is grown without irrigation. Major pest problems are lygus bugs, cotton fleahopper, bollworm, Verticillium wilt, and root-knot nematodes.

Development and Growth Requirements of the Cotton Plant

Understanding crop biology is essential to successful pest management. Without knowing how pests and management practices interact to affect plants, you cannot accurately evaluate pest injury and you may confuse it with other kinds of stress. IPM methods are chosen for their effect on the crop as a system, rather than solely for their impact on a specific pest.

Development

Cotton growth follows a definite pattern; roots, leaves, stems, and fruit are produced in a fixed sequence, although the timing of events is influenced by variety, temperature, spacing, irrigation, pest damage, and other factors. The pattern described here applies to both upland cotton, *Gossypium hirsutum*, and American Pima cotton,

G. *barbadense*, except in a few cases where important differences are noted. Although both species are perennials, they are usually grown as annual crops.

Germination

A mature cotton seed contains an embryo seedling and a supply of oils and proteins to provide energy for germination. Growth begins as the seed absorbs moisture from the soil, mostly through specialized cells at its large end. At germination, the primary root forces its way through the small end of the seed and pushes down into the soil (Figure 2). The hypocotyl—the stem portion between the root and the cotyledons—elongates and arches as it approaches the soil surface. By the time the hypocotyl pushes through the surface, the cotyledons have already

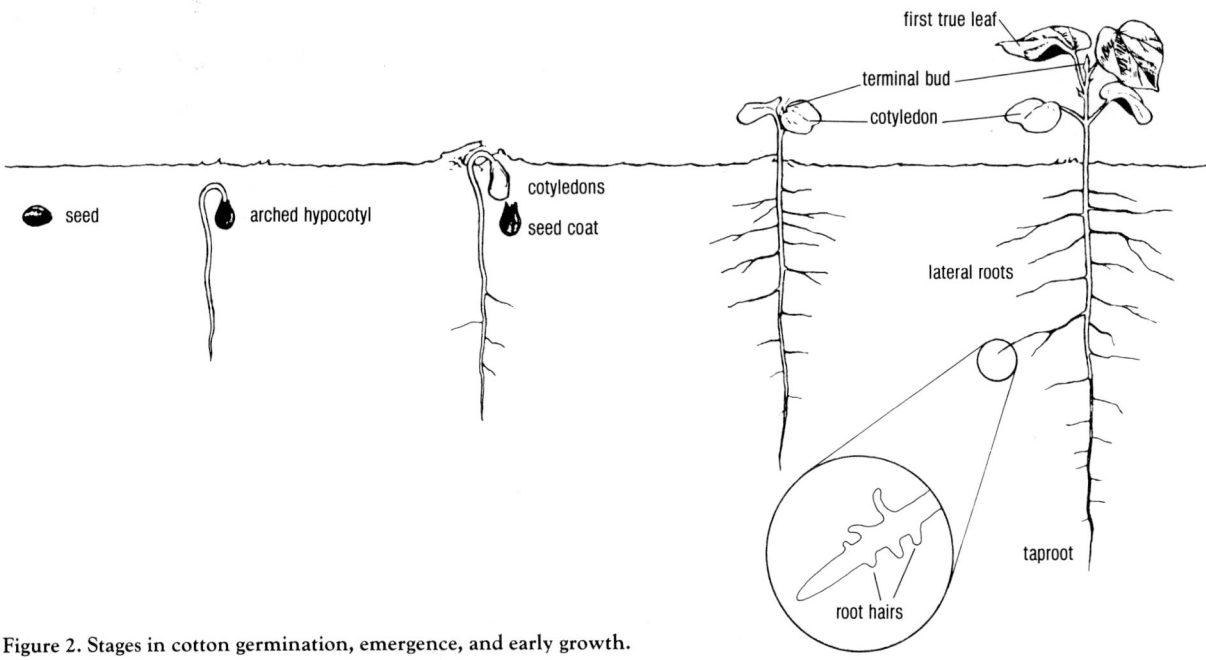

seed

arched hypocotyl

cotyledons

seed coat

first true leaf

terminal bud

cotyledon

lateral roots

taproot

root hairs

Figure 2. Stages in cotton germination, emergence, and early growth.

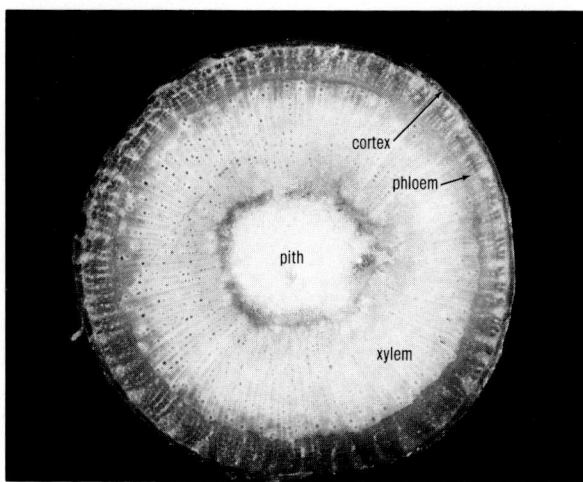

Figure 3. Cross section of cotton stem showing the positions of the xylem and the phloem, the two elements of the vascular system. The xylem carries water and nutrients from the roots through the rest of the plant, and the phloem distributes the products of photosynthesis.

expanded enough to begin producing energy for early growth.

Cotyledons in most cotton varieties have numerous tiny, black glands that produce gossypol, a substance poisonous to humans and to some animals, including pigs and chickens. Gossypol glands later appear in stems, true leaves, bolls, and other plant parts. Gossypol helps protect the plant from damage by vertebrates such as rabbits and rodents, but it also reduces the value of cottonseed for food and feed. "Glandless" varieties lack gossypol glands in the seeds or in the entire plant.

Roots

The primary root becomes the taproot, normally growing straight down to a depth of 6 feet (2 m) or more. By the time the seedling emerges, the taproot has penetrated about 10 inches (25 cm) into the soil and is starting to develop lateral branches. Fine root hairs at the tips of the laterals absorb water and nutrients that are then distributed through the plant in the vascular system (Figure 3). Most laterals are usually in the upper 2 feet (60 cm) of soil.

If the taproot encounters poorly drained soil, runs into a compacted layer, or is injured by root-knot nematodes or seedling disease, the plant may develop only a shallow system of laterals. During the first 3 or 4 days of growth the developing taproot is especially vulnerable to injury due to cold soil or excess moisture.

The root system is well established by the time of flowering. Because there is little root growth after fruit set begins, a plant with a shallow root system remains vulnerable to water stress throughout the season.

Main Stem and Leaves

As the main stem grows from a terminal bud, it develops a series of nodes, where leaves and branches are attached; the sections between them are called internodes (Figure 4). Plant height depends on the number and length of the internodes. Each node above the first, where the cotyledons are attached, bears a leaf and usually has one or two branches. Leaves and branches are arranged in a spiral along the stem; this arrangement minimizes shading of lower leaves by upper ones (Figure 5).

Branches

Branches develop from buds in the axils at the bases of mainstem leaves. Fruiting branches, which bear most of the plant's fruit, grow as a series of segments, each developing from an axillary bud on the previous segment. This growth pattern gives fruiting branches a slightly zigzag appearance. Vegetative branches, which may give rise to secondary fruiting branches, mainly bear leaves; they grow from a terminal bud in the same way as the main stem.

Western cotton varieties usually have the first fruiting branch at the sixth or seventh node above the cotyledons; lower branches are vegetative. The first fruiting branch may not appear until the eighth node or higher, however, if plant density is high. Appearance of the first fruiting branch may also be delayed if high nighttime temperatures occur when the first true leaf is expanding. Early applications of methyl parathion can cause a similar delay because of injury to the bud.

Fruiting Structures

The cotton fruiting structure begins as a flower bud, or "square." After flowering, it becomes a true fruit called a boll. Squares consist of a series of concentric whorls. The outer whorl—the only one visible until the flower is nearly ready to open—consists of three bracts. The bracts normally are clasped tightly around the square, but when a square has been injured by insects or other causes, the bracts often flare open and turn yellow before the square drops from the plant.

Just inside the bracts, at the base of the flower, is the calyx. When the square is very small, the calyx completely encloses the petals, but the calyx is pushed open as the square develops. On a flower, the calyx is a small green ring around the base of the five petals.

At the center of the square or flower is the upright, cylindrical pistil (Figure 6). Surrounding it are numerous stamens with pollen-producing anthers at their tips. The enlarged, conical ovary at the base of the pistil consists of two or more seed chambers, called locules by botanists but known to growers as "locks." Most upland cotton flowers have four or five locks; Pima cottons usually have three.

Each lock contains eight to twelve ovules that develop into seeds if they are fertilized. In upland cotton, seven to nine ovules usually mature in each lock; in Pima cottons, five to seven ovules mature.

Cotton flowers usually remain open for one day. The petals of upland cottons, white on the day they open, turn red as they close the next day. Petals of Pima cottons are yellow with a purple spot near the base; they do not flare open as widely as petals of upland flowers. Like upland flowers, Pima flowers turn dark in a day.

Like many plants, cotton produces nectar that attracts insects, but cotton is unusual in that the nectaries are not limited to the flower. The nectaries within the flower are inside the calyx, and the nectar accumulates at the base of the petals. Most varieties also have nectaries at the base of the bracts and on the lower sides of the main leaf veins, and there are numerous microscopic nectaries on petioles and flower stalks.

Because a square is produced at every node on fruiting branches, each node has the potential to bear a flower and eventually a boll if the fruiting structure is not lost due to stress or pest damage. While plants are growing vigorously during the first part of the season, there is a period of about 3 days between nodes on the main stem

Figure 4. Arrangement of branches and fruiting positions on a cotton plant.

Figure 5. Arrangement of mainstem cotton leaves as seen from above. The newest partially unfurled leaf, number 1, is the youngest leaf with part of its upper surface exposed. Young leaves are glossy, but fully expanded leaves such as number 4 or 5 have a dull surface. You will need to identify specific leaves when taking petiole samples for nutrient analysis and in sampling for bollworm, tobacco budworm, and spider mites.

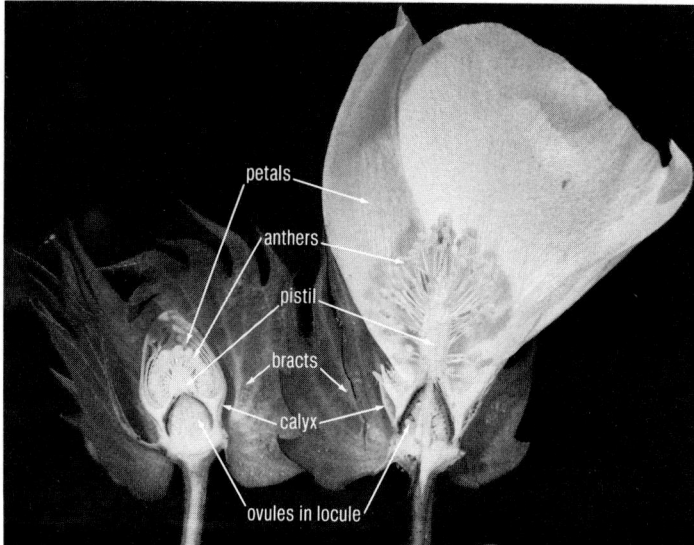

Figure 6. Anatomy of a cotton square and flower.

and about 6 days between flowers at successive nodes on the same branch. These intervals are longer during cool weather and they also increase later in the season, when growth slows due to fruit set.

Boll, Seed, and Lint Development

Fruit development requires pollination and fertilization. Cotton flowers are usually self-pollinated; pollen that fertilizes the ovules comes from the anthers of the same flower. The anthers burst when mature, releasing pollen which contacts the nearby stigma, a sticky portion at the tip of the pistil. Cotton pollen is relatively heavy, so it is not readily carried to other flowers by wind. Bees may carry pollen from one flower to another, but pollination by bees is seldom important under western conditions. Pollination is usually completed the morning the flower opens.

Once a pollen grain reaches the stigma, it germinates and produces a pollen tube that extends through the pistil into the ovary. Sperms then travel down the tube to fertilize an ovule. One pollen grain is needed to fertilize each ovule. Usually a few ovules in each lock are not fertilized; these remain as contaminants called "motes" in the lint. If too few of the ovules are fertilized, the fruit will drop within 10 days after flowering.

After fertilization, the bolls and their developing seeds expand rapidly. Inside the seed, the embryo seedling is fully formed about 5 weeks after fertilization. The oils and proteins that will provide the energy for germination accumulate mostly during the latter part of the seed's development.

While the embryo takes shape inside the seed, cotton fibers develop from the outermost layer of cells on the seed coat. There are two kinds: the lint fibers and the much shorter fuzz fibers or linters that remain attached to the seed after ginning. Each fiber originates as a narrow projection from a single cell and takes about 18 to 21 days to reach its final length. Near the end of this period, a series of cellulose layers is deposited on the inner surface of the fiber wall, forming a strong, laminated structure. At maturity, the lint dries and the fibers become twisted in a way that makes them especially suitable for spinning. As bolls dry, they split and flare open, exposing lint and seeds. Before ginning, the lint and seeds are called seed cotton.

Growth Requirements

A cotton plant, like a factory, needs a supply of raw materials and a source of energy. The raw materials for plant growth are carbon dioxide (CO_2), oxygen, water, and mineral nutrients. The energy of sunlight is captured by chlorophyll and other chemical structures in green parts of the plant. Like other green plants, cotton acquires

CO_2 from the air and uses the trapped energy of sunlight to build it into sugars and other compounds that serve as the plant's energy supply and as building blocks for growth. This process is called photosynthesis.

Most water used by plants passes from the roots up through the vascular system and evaporates through leaf pores, the stomata. Only a small amount actually remains in plant tissue. Evaporation from leaves—transpiration—occurs because stomata must remain open to expose a moist surface that can absorb CO_2 from the air. The flow of water from roots to leaves also supplies water used in photosynthesis, carries nutrients throughout the plant, and serves as a means of cooling. Transpiration rates increase when temperature is high and humidity is low, so these factors influence the amount of water needed.

Plants need mineral nutrients from the soil to build proteins, chlorophyll, and other components. The nutrients needed in the largest amounts are nitrogen, phosphorus, potassium, calcium, magnesium, and sulfur. Nutrients required only in very small amounts but still essential are iron, boron, manganese, zinc, molybdenum, copper, and chlorine.

Oxygen has a dual role in plant growth; it is a byproduct of photosynthesis and is released into the air, and plants also require it to metabolize sugars and other compounds in respiration. The availability of oxygen is critical for roots as well as for aboveground plant parts, and roots may be injured in waterlogged or poorly aerated soil.

Growth Rate

As long as growth requirements are met, temperature is the main factor that determines how fast a plant grows (Figure 7). Growth begins only when the temperature is above a certain level, the developmental threshold. As temperature increases above the threshold, the growth rate increases to a maximum. Beyond the maximum, the growth rate declines and when temperature reaches an upper limit, it stops. The same general relation between growth and temperature governs development of insects, bacteria, nematodes, fungi, and most other organisms except warm-blooded vertebrates.

The relationship between growth and temperature is complex. In most cases, the developmental threshold is not fixed, but varies according to the stage of growth. In cotton, for example, the thresholds for germination and lint development are higher than the threshold for vegetative growth. It is not usually necessary, however, to take these differences into account for purposes of pest management.

Because temperature is so important in controlling growth rates, it is misleading to describe development in terms of time alone. For example, cotton is often described as requiring 50 days from emergence to first square, but

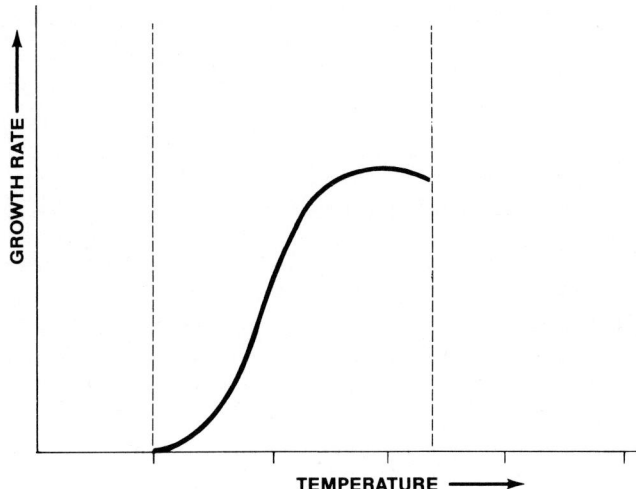

Figure 7. Approximate relation between temperature and growth rate. As temperature increases above the developmental threshold (dashed line at left), growth rate increases to a maximum. Beyond the maximum, growth again slows as temperature approaches the upper limit for growth (dashed line at right).

the time can be much longer if the weather is abnormally cool. The same kind of error could result from assuming that a certain pest always takes 3 weeks for a generation; depending on temperature, it may actually require 2 to 5 weeks. To monitor crop development and predict pest behavior, pest management uses a system that takes into account the accumulation of heat with passing time. This system is based on the *degree-day* (abbreviated °D).

One degree-day is the amount of heat that accumulates during a 24-hour period when the average temperature is one degree above the developmental threshold. Degree-days may be based on either Celsius or Fahrenheit temperatures, as long as the same scale is used consistently. The term *heat unit* is sometimes used instead of degree-day.

The simplest way to estimate the number of degree-days accumulating on a given day is to use the formula:

$$\frac{\text{Daily High} + \text{Daily Low}}{2} - \text{Developmental threshold} = \text{Degree-days}$$

For example, with a threshold of 60° F, one day in which the high was 90° F and the low was 70° F would be equivalent to 20 °D:

$$\frac{90°\,F + 70°\,F}{2} - 60°\,F = 20\,°D.$$

This formula is not very accurate when the daily low is below the developmental threshold. More accurate methods of calculating degree-days are presented in *Degree-Days: The Calculation and Use of Heat Units in Pest Management* (References).

Crop Models

Mathematical descriptions called *models* use the accumulation of degree-days, starting from a specific time such as planting, to predict the timing of events in cotton development. Such models can serve as a standard for evaluating the crop's progress each season. By keeping a record of degree-day accumulation starting at planting, you can judge when to expect various growth and fruiting stages, regardless of when the crop was planted. If the crop lags behind the standard, this will alert you to problems that you may be able to correct through adjustments in irrigation, fertilization, or other management practices.

Degree-day models can also correlate crop development with pest life cycles. In the desert valleys, for example, it is possible to predict the emergence of pink bollworm moths each generation. By comparing the emergence peak to the crop's fruiting cycle, you can identify periods when the crop is most vulnerable and when monitoring is most critical.

Several models are used widely in the Western Region. Some apply only to cotton development, while others are used both for the crop and for certain insect pests. All are based on degree-days, although they differ in the temperatures used as developmental thresholds. For example, one model uses 60° F as the threshold for cotton growth; another uses 53.4° F. Also, some models include a maximum temperature and others do not. The reason for these differences is that the models have been developed independently, often for different purposes, and they have been applied in different areas.

For routine pest management, differences among various cotton models are minor. The important point is to use a degree-day system rather than to rely on calendar dates for scheduling management activities. You must, however, use the same model consistently, because it is not possible to add degree-day figures based on different thresholds. Methods of obtaining temperature data and calculating degree-day values are discussed in the section on monitoring. Table 1 lists degree-day values for various stages of cotton development, as calculated from commonly used thresholds.

Energy Resources and Yield

Good cotton management enables the plant to channel as much energy as possible into harvestable seed cotton, within the limitations imposed by the soil, length of the season, and costs of production. Producing a high yield at reasonable cost requires the right balance between the *supply* of energy within the plant and the *demand* placed on energy resources.

A cotton plant's energy supply consists of carbohydrates—the sugars and other compounds manufactured in photosynthesis. Energy from carbohydrates is used for two basic functions, respiration and growth, that constitute the demand on the plant's energy resources.

Respiration, the process that supplies energy needed to maintain living cells, always has first priority in its demand for energy, since the plant would die without it. Energy remaining after respiration needs are met is available for growth. However, plants cannot support the growth of roots, stems, leaves, and fruit at maximum rates all at

Table 1. Degree-Days Required for Cotton to Reach Various Stages of Development. Figures in each line are degree-days calculated from developmental thresholds at left.

Base Temperatures	DEGREE-DAYS FROM PLANTING REQUIRED TO REACH:				
	Emergence	First Square	First Open Flower	First Boll Susceptible to Pink Bollworm	First Open Boll
Developmental Threshold = 60° F (San Joaquin Valley Acalas)	50	500	740	830	1800
Developmental Threshold = 53.4° F Upper Temperature = 94° F (San Joaquin Valley Acalas)	60	600	1550	*	2800
Developmental Threshold = 60° F (Desert valley Deltapines)	*	550	920	1040	1990
Developmental Threshold = 55° F Upper Temperature = 86° F (Desert valley upland varieties)	120	700	1190	1570	*
Developmental Threshold = 55° F (New Mexico Acala 1517s)	100	760	1200	2100	2380

*data not available

the same time, so they allocate energy to the various kinds of growth according to a system of priorities (Figure 8).

Early in development, vegetative growty—the growth of roots, stems, and leaves—has first priority for energy after respiration. During this period, the plant builds the vegetative framework needed to support the growth of fruit later. Because the production of squares is linked to vegetative growth, the plant also accumulates a stock of potential fruit.

Once a plant begins to set bolls, maturation of bolls takes precedence over vegetative growth and production of new squares. A boll begins to place significant demand on the plant's energy supply about 10 days after flowering. Demand increases steadily for another 25 days or so, then declines to zero by about 42 days after flowering. When a certain number of developing bolls have accumulated, they consume most of the energy remaining after respiration needs are met. Vegetative growth then slows or stops (Figure 9).

The degree to which growth is interrupted by fruit set depends on variety and growing conditions. Varieties in which the interruption is pronounced are often called "determinate," although in the botanical sense they are indeterminate. Pima cottons build up their boll load more gradually than do most upland varieties and tend to continue growing and producing fruit at a more or less constant rate throughout the season.

If the season is long enough, vegetative growth will resume after most early bolls have completed development and no longer require energy. In the San Joaquin Valley and in New Mexico, there is usually only one period of rapid vegetative growth followed by one fruit maturation period (Figure 10). In the desert valleys, however, the season is long enough for two cycles. Bolls produced during the second cycle are called the "top crop" by desert growers. The period between the two cycles, when the production of new squares virtually stops, is known as "cutout."

Carbohydrate Stress

Stress is any condition in which growth is limited by the supply of an essential factor such as water or a nutrient. Plants in which vegetative growth has slowed because of the energy demand imposed by boll development are in a state of *carbohydrate stress*; the factor in short supply is carbohydrates. Carbohydrate stress usually begins around peak bloom.

The timing of carbohydrate stress has a major influence on yield. For maximum yield, plants must grow enough that their leaves can supply energy for as many bolls as will have time to mature during the season. If carbohydrate stress stops growth too early, plants produce fewer squares than the maximum that could have matured. If stress is delayed too long, there will be excessive vegetative growth which, like excess capacity in a

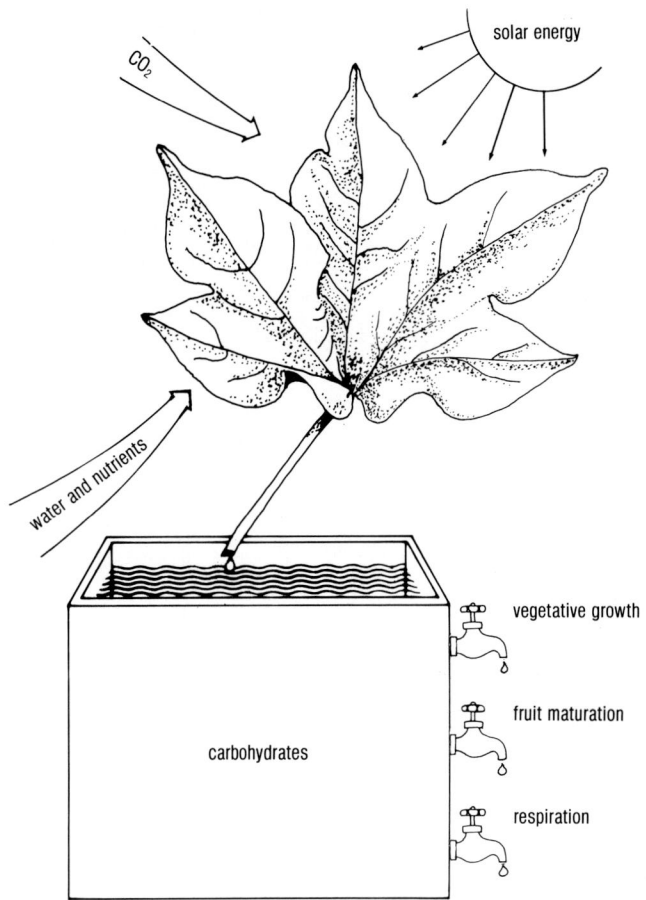

Figure 8. Carbohydrates produced in photosynthesis form an energy pool that supports all plant functions. As demand drains energy, vegetative growth is the first function to shut off; fruit maturation is next, and respiration is last.

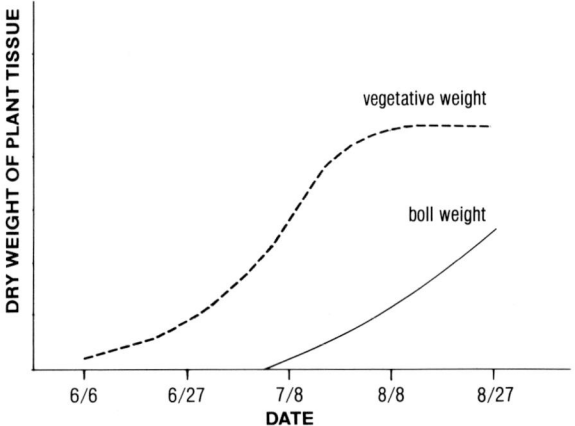

Figure 9. As boll load increases, vegetative growth slows. Boll load is measured as total dry weight of all bolls on the plant. Vegetative growth is measured as dry weight of leaves, petioles, and stems. This figure is based on data from Acala cotton in the San Joaquin Valley.

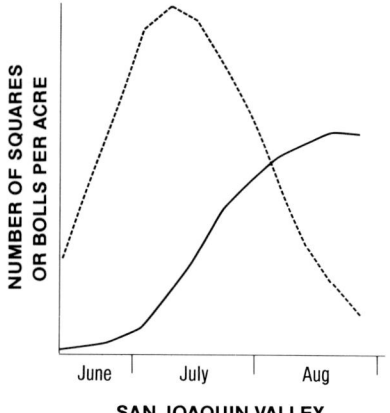

SAN JOAQUIN VALLEY

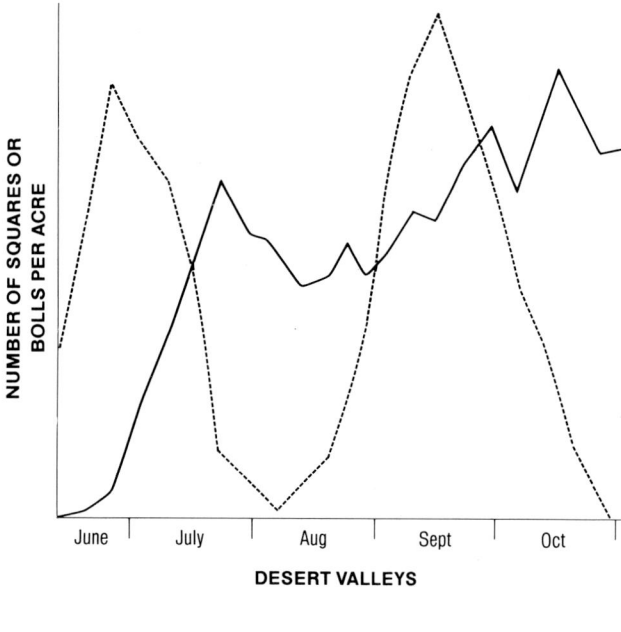

DESERT VALLEYS

——————— bolls

------------- squares

Figure 10. In the cotton fruiting cycle, production of squares reaches a peak, then declines as mature bolls begin to accumulate on the plant. In the low desert valleys, the season is long enough to allow a second cycle of square production after bolls of the first cycle have matured and no longer require energy.

factory, is wasted if there is not enough time to fill the demand. A large plant with numerous squares may produce fewer mature bolls than a much smaller plant if there is too little time in the season for all the fruit to mature.

Cultural practices and certain kinds of pest injury can affect the balance between supply and demand within plants. Plant density is an important factor because plants that are closer to each other compete more for sunlight, water, and nutrients. When growth is limited by competition, each plant's energy supply is smaller, so it takes fewer bolls to consume the entire supply and to slow further growth.

Water and nitrogen also affect the supply/demand relationship. While an adequate supply of both is needed to develop the vegetative framework that supports fruit growth, too much water or nitrogen may keep plants in the vegetative growth phase too long. As a result, plants may become "rank" with excessive foliage and stems; rank growth uses energy that would otherwise contribute to yield. Also, upper leaves on rank plants often shade lower ones, reducing their ability to supply energy to adjacent bolls. Foliage produced late in the season at the top of the plant or at the tips of long branches contributes little, because bolls in these positions seldom have time to mature.

Shedding of Fruiting Structures

Cotton plants produce far more fruiting structures than they can support to maturity with the available energy resources, and they shed the excess. This normal shedding due to carbohydrate stress thins the fruit load. The rate of shedding increases rapidly soon after the boll load begins to build up (Figure 11). Of the squares produced early in the fruiting period, 40 to 50% are eventually shed; of those produced late in the fruiting period, up to 90% are shed. Plants shed both squares and bolls in response to carbohydrate stress. Squares are usually shed soon after the pinhead stage; bolls are usually lost at 10 days old or less. Plants will shed increasingly older squares and bolls, however, if stress is severe due to disease, excessive leaf damage by mites or insects, or other factors. Upland varieties usually shed 60 to 70% of the fruiting structures they produce. Pima cottons produce fewer fruiting structures, but retain more of them.

Squares and bolls that are close to the plant's energy resources and those that appear early in the fruiting cycle have the best chance of maturing. The energy that supports development of a square or boll comes mainly from the closest leaves. A boll at the first node on a fruiting branch receives most of its energy from the adjacent leaf and from the mainstem leaf at the base of the branch. Bolls at nodes beyond the first one receive less energy from the mainstem leaf; because their adjacent leaves are smaller, they have a smaller total energy supply.

These differences are reflected in the proportion of yield that comes from different fruiting positions (Figure 12). In most upland cottons at normal plant densities, about 75% of total yield comes from bolls at the first node on fruiting branches; about 15% comes from the second position, and only about 10% comes from positions beyond the second. Acala 1517 cottons grown in New Mexico normally produce 85 to 90% of their yield at the first position. At high plant densities, the proportion of yield that comes from the first position is even greater.

Other stresses may cause plants to shed fruiting structures at different stages of development. Water stress causes loss of small squares; the more severe the stress, the larger the squares affected. If severe water stress occurs while plants have a boll load, bolls as well as squares may be shed.

Effect of Pest Injury on Energy Balance

Many kinds of pest injury can best be understood in terms of their effect on the division of energy resources in the plant. Verticillium wilt, for example, gradually chokes off a plant's energy supply by reducing the plant's ability to distribute the products of photosynthesis through the vascular system. The first plant activity to be affected is vegetative growth; it stops even before external symptoms appear. As the disease progresses and the energy supply declines, the maturation of existing bolls may also stop. With severe disease, respiration may eventually stop and the plant will die.

Lygus bugs can dramatically affect energy allocation by destroying small squares early in the season. If the bugs remove too many squares, there will not be enough bolls later to channel energy away from vegetative growth, and plants may become tall and rank. The loss of some early squares is not necessarily detrimental, however, as long as the season is not so short that time becomes a limiting factor. With fewer early squares and a delay in the energy demand imposed by developing bolls, plants may develop a stronger root system and larger vegetative framework that can support more bolls later. Management may be more difficult, however, in crops where loss of early squares produces a delayed, but more concentrated, fruit maturation period.

Root-knot nematodes have a double effect on energy resources. They increase demand by stimulating the growth of root galls and by consuming products of photosynthesis directly. At the same time, they limit the energy supply by interfering with the flow of water and nutrients through infected roots.

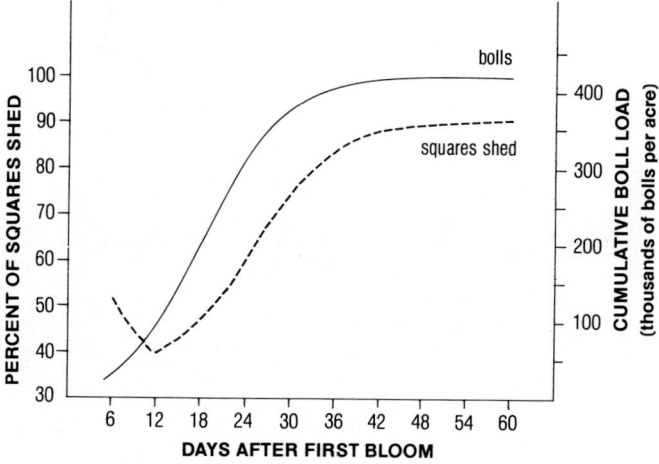

Figure 11. Shedding of squares increases as boll load builds up. This graph is based on studies of Acala cottons in the San Joaquin Valley.

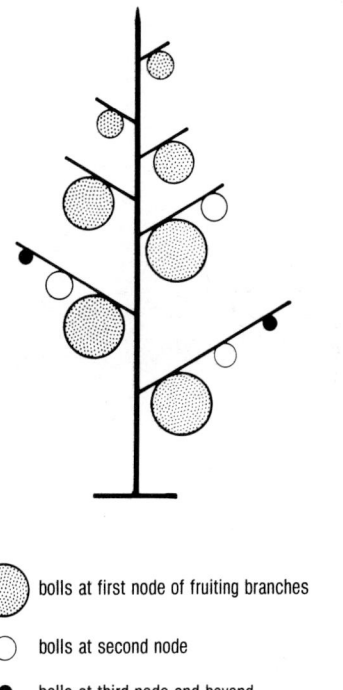

Figure 12. Most cotton yield comes from bolls close to the plant's main energy resources.

Managing Pests in Cotton

Because climate, soil type, cropping history, cultural practices, crop variety, and the nature of surrounding land vary from field to field and from season to season, no single management program is suitable for all cotton crops. Market conditions also affect management options because they determine the value of the crop. Regardless of conditions, however, four components are essential to any IPM program:

- accurate pest identification;
- field monitoring;
- control action guidelines;
- effective prevention and control methods, including use of appropriate pesticides when needed.

The purpose of this chapter is to help you incorporate these components in a program tailored to the specific needs of each field.

Pest Identification

Because most pest management tools, including pesticides, are effective only against a certain range of species, you must know which pests are present in a particular field and which are likely to appear. Different methods or pesticides may be needed even for closely related species.

Use the descriptions and photos in this manual to identify pests commonly found in western cotton. Check the References for other sources of information. Remember that some pest problems can be diagnosed reliably only by experienced professionals; don't hesitate to seek technical help if you are not sure of an identification. Farm advisors and county Extension agents can help you identify pests and can direct you to specialists if necessary.

Monitoring

Monitoring provides information needed to make management decisions. It includes keeping records of weather, crop development, and management practices, as well as evaluating pest infestation levels. Seasonal monitoring activities are summarized in Figure 13.

Weather Data and Degree-Days

Because weather influences development of cotton and its pests, monitoring requires a reliable source of weather data. Daily high and low temperatures are needed to determine the accumulation of degree-days through the season. Evapotranspiration data (page 26) may be useful for scheduling irrigations, and weather forecasts are essential for scheduling planting and other operations. Many newspapers and radio stations in agricultural areas report such information, and the National Weather Service broadcasts weather information over very high frequency (VHF) channels on FM radio.

Because temperature is affected by local variations in terrain, vegetation cover, elevation, and other conditions, regional weather information may not reflect the situation in your field. The best source of temperature data is a recording maximum/minimum thermometer or similar instrument placed as near as possible to the field you are managing. Follow the manufacturer's recommendations in setting up weather instruments; in general, they should be 5 feet from the ground, sheltered from direct sun, and placed away from paved surfaces that radiate heat.

Given daily high and low temperatures, you can determine how many degree-days accumulate each day. By adding the daily figures, you can obtain a total for longer periods or for the whole season. The easiest way to obtain degree-day values is to use tables such as those in the Appendix. Methods of calculating degree-days from temperature data are discussed in *Degree-Days: The Calculation and Use of Heat Units in Pest Management* (References).

Regardless of whether you obtain degree-day values from a table or calculate them yourself, remember that the results can be no more precise than the temperature data on which they are based. The further away the source of the temperature data is, the less accurate the results.

Monitoring Pest Activity

Monitoring during the growing season is devoted mostly to checking insect and mite populations. Visit each field at least once a week; during critical periods of crop development or when a population is approaching a treatment threshold, visit every 2 or 3 days. In each growing area, the emphasis is on two or three pest species that occur regularly and that threaten yields or quality. For each one, sampling is needed only during a specific period in crop development, so you seldom will need to perform more than two procedures during each visit. Work sampling methods into a single routine so that you do not have to make a special pass through the field for each one.

Other pest monitoring activities are needed only once or a few times each season. Start before planting by conducting a weed survey. Repeat the survey once or twice during the growing season and again at harvest. In fields with sandy soil or a history of nematode injury, collect soil samples for laboratory extraction of root-knot nematodes. In areas where Verticillium wilt is prevalent, you may also need to collect soil samples for analysis of the inoculum level.

Soil, Water, and Tissue Testing

Contact a reliable laboratory for testing soil, water, and plant tissue. Preplant soil analysis helps determine the need for nitrogen and other nutrients, and soil tests can also measure salinity, soil reaction (pH), organic matter content, and other factors that affect crop growth. Annual testing may be needed for nitrogen, but other tests are not usually needed more often than every 2 or 3 years.

Occasional testing of irrigation water helps prevent harmful increases in salinity, especially when the water comes from wells or where surface water is known to be high in salts. Tests should measure the sodium level and the concentration of total salts.

Analysis of plant tissue is the best way to monitor the crop's nutrient status during the growing season. A typical program requires collecting samples of petioles and/or leaf blades for nutrient testing three or four times each season.

Keeping Records

Knowing what happened last year can make a lot of management decisions easier. Keep a file, notebook, or computer record of the following for each field:

- weekly monitoring reports;
- weed surveys;
- records of pesticide applications, including names of materials, rates, and dates of application;
- laboratory reports;
- agronomic information, including crops and varieties planted, planting and harvest dates, and yields.

In addition to maintaining records that apply to specific fields, keep a file of local weather data, including charts of accumulated degree-days.

Control Action Guidelines

Control action guidelines indicate when management actions, including pesticide applications, are needed to avoid losses to pests. Guidelines for insect pests are generally numerical thresholds based on specific sampling techniques; they are intended to reflect the population level that will cause economic damage if left unchecked.

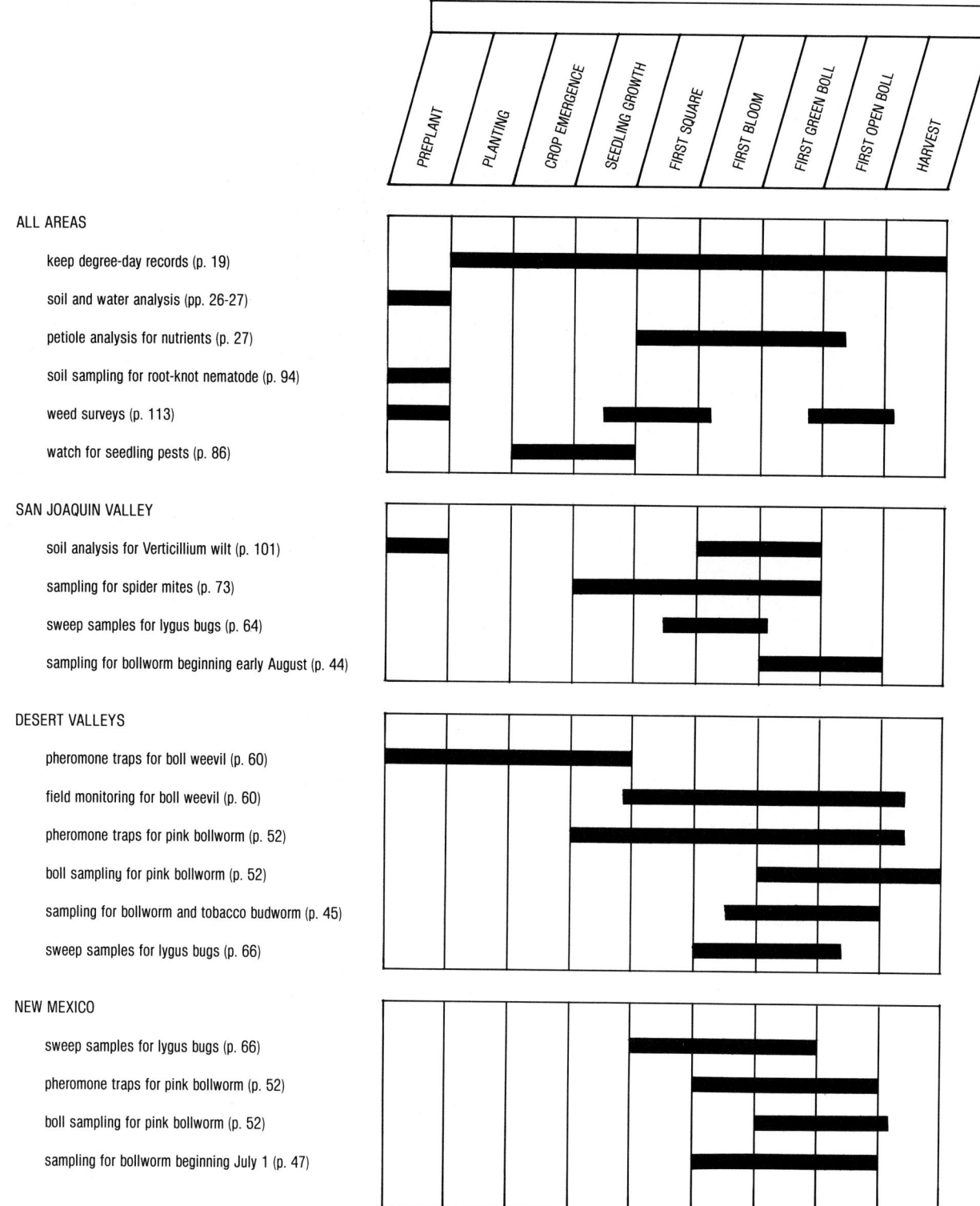

Figure 13. Summary of monitoring practices needed each season in different parts of the Western Region.

Numerical guidelines are also available for root-knot nematodes and Verticillium wilt in the San Joaquin Valley. Guidelines for other pests, including most pathogens and weeds, are usually based on the history of a field or region, the stage of crop development, weather conditions, and other observations.

Control action guidelines often change as new varieties and cultural practices are introduced and as new information on pests becomes available. The guidelines presented in this manual may be revised as more research is completed on cotton pest management.

Management Methods

The ideal IPM program protects the crop while interfering as little as possible with long-term maintenance of the production system. The cheapest, most reliable way to deal with pest problems is to anticipate and avoid them. When pesticides are needed, choose materials and application methods that control pests effectively with a minimum of harmful side effects.

Biological Control

Biological control is any activity of a parasite, predator, or pathogen that keeps a pest population lower than it would be otherwise. Its role is best understood in the case of insects; without natural enemies, many pests normally of little importance would increase to levels that would make cotton farming economically impossible. In the San Joaquin Valley, for example, natural enemies and diseases limit populations of bollworms, beet armyworms, loopers, and other potentially destructive caterpillars that feed on cotton. The absence of effective natural enemies is one factor that makes the pink bollworm a key pest in the desert valleys.

To gain the greatest benefit from biological control of insect pests, choose insecticides, rates, and application methods that have a minimum impact on natural enemies. Natural enemies that affect nematodes, weeds, and fungi are being studied, but as yet no practices are recommended for improving biological control of these pests.

Resistant Varieties

Varieties tolerant or resistant to disease can provide long-term, economical protection from conditions that otherwise could inflict severe losses every season. Compared to earlier varieties, Acala cottons grown in the San Joaquin Valley and New Mexico have significant tolerance to Verticillium wilt, and this tolerance helps to keep cotton profitable. Varieties resistant to bacterial blight are valuable in New Mexico, where this disease occurs regularly. Some experimental cottons have significant

Figure 14. The parasitic wasp, *Hyposoter exiguae,* lays an egg in a young beet armyworm. Natural enemies such as this play an important role in control of many cotton pests.

resistance to root-knot nematodes, but resistance is not yet available in commercial varieties.

Some cotton varieties have characteristics that make them less susceptible than others to insect damage. Nectariless cottons lack the extrafloral nectaries that provide food for many insects, including lygus bugs and pink bollworm moths. Smooth-leaved varieties are less attractive to egg-laying bollworm and tobacco budworm moths, while highly pubescent varieties discourage feeding and egg laying by cotton fleahoppers. Some cottons, including common Pima varieties, have high concentrations of compounds that inhibit feeding by bollworms and budworms. No single characteristic by itself provides enough resistance to major pests to make this a significant factor in selecting varieties. However, if plant breeders successfully combine two or more resistance traits in a single variety with desirable agronomic qualities, improved resistance to insects may be available in the future.

A varietal characteristic important to pest management in other parts of the U.S. cotton belt is earliness of fruiting. Early varieties are a major component of management strategies for pink bollworm, boll weevil, and tobacco budworm in some areas. Although they are not widely planted in the Western Region, early varieties, in combination with short-season culture, are available as a management option if late-season pests cannot be controlled by other means.

Cultural Practices

Many cultural practices, including seedbed preparation, planting, irrigation, and field sanitation, have a significant impact on pest damage. Even when you cannot choose cultural methods solely for their effect on pest management, it is important to understand their impact on pests so that you will know what to expect.

Try to manage the crop so that it will set fruit when plants are best able to support the fruit load and while there is ample time for fruit to mature. In the San Joaquin Valley, this means protecting squares from damaging populations of lygus bugs during the critical third through sixth weeks of squaring, and avoiding water stress during the flowering period. In short-season areas, such as New Mexico, the entire squaring period is critical because there is not enough time for the crop to replace lost squares.

Even in the desert valleys, where the season is long enough for two fruiting cycles, it is more efficient to get the most out of the first cycle than to rely on the later "top crop" for maximum yield. In well-managed cotton, the top crop is seldom more than 20% of the total yield. The cost of producing the top crop can be higher than the value of the extra yield, due partly to the need for insecticides to control pink bollworm, tobacco budworm, and other pests that build up late in the season. Extra water and fertilizer are also needed. There may also be extra costs the following season, when earlier control measures may be needed against high overwintering populations of pink bollworm or boll weevil.

Planting. A successful cotton crop begins with high quality, vigorous seed. When possible, use seed that has been rated in a "cold test"—a germination test in which seed samples are held for 6 days at 65° F (18° C). The cold test is more reliable than the traditional germination test in predicting percent germination, because the lower temperature is closer to normal field conditions at planting.

Fungicide seed treatments greatly reduce losses due to seedling diseases. Where the risk of seedling disease is high, as in fields with a history of stand reduction and in fields planted while the soil is cool and wet, in-furrow application of granular fungicide can provide extra protection. Acid-delinted seed is essential for preventing bacterial blight, especially in New Mexico, where the disease is most common.

Insecticidal seed treatments and application of insecticides to the soil at planting are not generally recommended. Insect pests of seedlings do not occur in damaging numbers regularly enough for treatments to be worthwhile and such treatments often destroy flower thrips, important predators of spider mite eggs.

Losses due to seedling pests and diseases are lessened and the chance of herbicide injury is reduced when soil temperature favors rapid germination and growth. Ideally, soil temperature at a depth of 8 inches (20 cm) should reach 60° F (15.5° C) for 3 days in a row before planting. Measure soil temperature at about 8 a.m., when it is close to the daily minimum. Because planting schedules are more often controlled by moisture conditions and the availability of equipment than by temperature, it may be necessary to plant when the soil is cool. It is essential, however, to make sure the weather forecast calls for at least 4 days of clear, warm weather after the intended planting date. Don't plant if rain or a cooling trend is expected.

In preirrigated fields, allow enough time for the beds to drain adequately before planting; wet soil takes too long to warm up and may delay emergence. Also, planters used in wet soil often create a shallow, compacted layer that restricts root growth and makes roots more susceptible to infection by soil fungi.

Keep the planting depth as shallow as you can while still allowing enough moisture for germination. The depth generally recommended is from 1 to 1½ inches (2.5 to 3.8 cm) into moist soil. Seeding should be slightly deeper in sandy, coarse-textured soils than in finer-textured soils, but the seed should never be planted deeper than 2 inches (5 cm). Check equipment before and during planting to make sure that all planter gangs are placing seed at the same depth and that firming wheels are tracking seed rows accurately.

Planting density under western conditions is usually from 20,000 to 60,000 plants per acre or about 1½ to 4½ plants per foot of row. Higher densities may be needed where growth is limited by salinity or other soil factors. In general, the optimum plant density is lower in fertile soils that produce large plants than in less fertile soils, and it is lower in late than in early plantings. Properly managed, a relatively high density can reduce losses from Verticillium wilt, as it tends to decrease the percentage of plants infected. Ask farm advisors or county agents for recommendations on plant density under local conditions.

Irrigation. Availability of soil water is a major factor that determines yield and the timing of crop maturity. Too little water will reduce yield; excess or poorly timed irrigation can delay maturity and favors unproductive lush growth. Irrigation can induce pest problems indirectly by lowering soil temperature, increasing humidity and lowering temperature in the canopy, and by increasing the attractiveness of plants to insects.

Efficient irrigation requires finding out how much *available water* the soil can hold. Available water is that portion of the soil water that can be withdrawn by plants

(Figure 15). During the growing season, irrigation is needed when a certain proportion of the available water, the *allowable depletion*, has been used. The allowable depletion in a particular field varies according to soil type, stage of crop growth, total amount of available water, weather conditions, and irrigation cost. For western cotton, it is usually from 50 to 80%.

Consult soils maps and use a soil probe to measure the crop's rooting depth and to identify any layers of different soil types. In deep, uniform soil, rooting depth may be 6 feet (2 m) or more by midseason, although most of the water used by the crop is drawn from the upper 3 feet (1 m). If there is a shallow hardpan or compacted layer, the crop will have a shallow root system and may need more frequent, but lighter, irrigations than it would in deeper soil.

Once you know the rooting depth, you can determine from the clay content and soil texture (Figure 16) how much available water the soil can hold. For a soil profile that includes layers of different soil types, calculate the available water separately for each one; then add them together to obtain the total available water in the rooting depth. Take salinity into account in estimating available water; if the soil or irrigation water contains high levels of salts, the ability of plants to withdraw water is reduced and the proportion of available soil water is lower.

Preirrigation and "Irrigating Up." Irrigation is usually needed at or before planting to provide water for germination and early growth. You can apply water before planting as a preirrigation, or "irrigate up"—that is, apply water shortly after planting. In either case, fill the soil reservoir early in the season so that you can keep up with the crop's water needs later.

In most soils, preirrigation is preferable to irrigating up; it allows filling the soil profile without cooling the soil too much after planting. In cool areas and in early plantings, irrigating up can lower soil temperature enough to slow germination and growth, thereby favoring seedling diseases. Except on sandy soils, it is difficult to fill the soil profile after planting without flooding and waterlogging. Also, some soils form a hard crust that inhibits seedling emergence if the seedrow is saturated just after planting.

The amount of water needed for preirrigation depends on rainfall and on the nature of the soil profile. Most of the Western Region seldom gets enough winter rain to fill the soil reservoir. Rainfall may be adequate, however, in wet years and in fields with a perched water table. Even with adequate rainfall, extra water may be needed to leach accumulated salts from the root zone.

Postplant Irrigations. Each postplant irrigation should bring the soil in the rooting depth back to field capacity. When moisture is available throughout the root zone, plants draw nutrients from a larger volume of soil and there is less

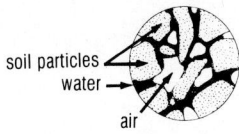

Soil is about half solid material by volume (large circle). The rest of the soil volume consists of pore spaces between soil particles; pore spaces hold varying proportions of air and water (small circle).

soil particles
water
air

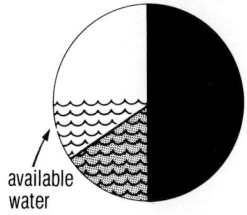

Just after a furrow or flood irrigation or heavy rain, the soil is **saturated**—the pore spaces are entirely filled with water.

After a field is allowed to drain following irrigation, the soil is at **field capacity**; in most soils, about half of the pore space is filled with water. About half of this water is **available water** that can be used by plants; the rest is unavailable because too much suction is needed to remove it from the pore spaces. The proportion of water available to plants is higher in clay soils and lower in sand.

available water

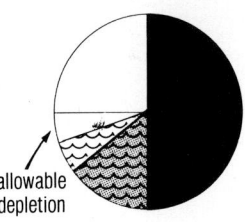

The **allowable depletion** is the proportion of the available water that can be used up before irrigation is needed. This proportion is set by the crop's tolerance for moisture stress under prevailing conditions and by the cost of supplying water.

allowable depletion

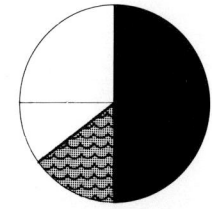

At the **wilting point**, all available water is gone; plants wilt and die unless water is added.

Figure 15. The soil reservoir.

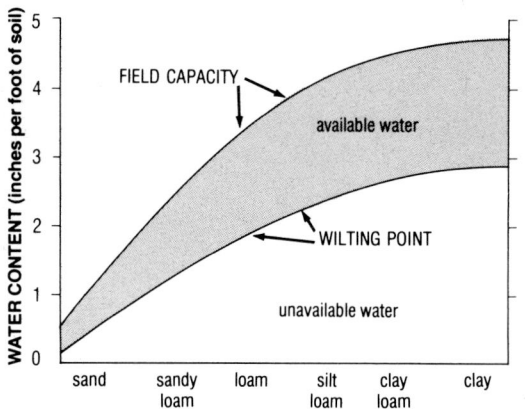

Figure 16. **The higher the percentage of clay in soil, the greater the water-holding capacity. The shaded area represents available water.**

chance of harmful stress if an irrigation is delayed later. If the root zone is shallow, if water penetration is slow, or if roots are damaged by nematodes, irrigate more often but apply less water each time.

The timing and amount of the first postplant irrigation depend on the water-holding capacity of the upper 2 feet (60 cm) of soil, where most of the crop's roots are before squaring. You must apply water early enough to avoid severe stress that will interrupt growth and delay fruiting. There is evidence, however, that a very mild degree of stress before first irrigation improves the retention of early bolls, maintaining or increasing yield while using less water. Irrigating too soon may promote Verticillium wilt by cooling the soil, and excessive early irrigation can make the crop more attractive to lygus bugs. Fields infested with nutsedge, on the other hand, may need early irrigation to compensate for competition from the weeds and to push vegetative growth so that the crop will be able to shade out the nutsedge as early as possible. A rule of thumb used in the San Joaquin Valley is that preirrigated cotton in sandy soils needs water by late May; cotton in clay-loam soils can usually wait until late June.

One or 2 inches (2.5 to 5 cm) of water is usually enough at the first irrigation. In furrow-irrigated fields, you may be able to control the amount more carefully by watering alternate furrows, although this technique will not work where soil cracks allow water to run from one furrow to the next.

The critical point in midseason irrigation is to avoid water stress during flowering, especially during the 3- or 4-week period around peak bloom. In the San Joaquin Valley, this critical period is usually from about July 10 to August 1. Plants stressed excessively during flowering will shed squares that they will not have time to replace. Under severe stress, they may also shed small bolls.

The effect of water stress on fruiting is often overlooked because the squares lost are usually very small. If you are not watching the crop closely, you may not notice the

loss of squares until about 3 weeks after the stress has occurred, when there may be a sudden decline in the number of flowers and small bolls. Since 3 weeks is the approximate interval between irrigations in many areas, an interruption in flowering due to earlier water stress often coincides with a later irrigation, and it is sometimes mistakenly attributed to the irrigation itself. Loss of small squares due to water stress has also been confused with lygus injury.

The last irrigation should provide only enough water for the crop to complete development of bolls that normally have time to mature. Irrigating too late in the season delays opening of mature bolls, favors boll rots, may increase lodging, and can make defoliation more difficult by promoting late growth and regrowth. In the deserts, late irrigations also favor high overwintering populations of pink bollworm and tobacco budworm.

The last irrigation must be later in sandy soils with low water-holding capacity than in fine-textured soils. In the San Joaquin Valley, the last irrigation can be as early as the first or second week of August in clay soils where the crop has a deep, healthy root system. In loamy soils, water is usually needed in late August, and in sandy soils it may still be needed in early September.

Irrigation Scheduling. Timing of irrigations during the growing season can be based on observations of water stress in plants and various measures of soil moisture. Where evapotranspiration (ET) data are available, a water budget is also helpful. A combination of two or more methods is usually best.

Regardless of the methods you use, always check soil moisture before applying water and estimate how much available water remains in the crop's rooting depth. Use a soil tube or shovel to take soil from the rooting depth at several points in each field. In furrow-irrigated fields, check a few places at the head end of the runs, some in the middle, and some at the tail end.

The allowable depletion for cotton during the flowering period is generally about 60% in fine-textured soils and up to 75% in sandy soils. Late in the season, before defoliation, depletion may be allowed to reach 80% in most soils, and even more in very sandy soils. Use Table 2 as a guide for judging the depletion level in soil taken from the root zone.

For an experienced observer, the condition of crop plants indicates their need for water. In a vigorously growing cotton plant, the upper portion of the main stem remains green, while the lower stem is reddish. The more the plant is stressed for water, the closer the reddish color is to the growing tip. Other symptoms of water stress include wilting, a bluish tinge on leaves, and an excessive number of flowers at the tops of plants.

Many San Joaquin Valley growers watch the red color of the main stem and apply water often enough to keep the upper 3 or 4 inches (7 to 10 cm) of the stem green.

Table 2. Judging Depletion of Soil Water by Feel and Appearance. Numbers in each column are inches of water needed to restore 1 foot of soil depth to field capacity when soil is in the condition indicated.

Coarse-Textured Soils	Inches of Water Needed	Medium-Textured Soils	Inches of Water Needed	Fine-Textured Soils	Inches of Water Needed
Soil looks and feels moist, forms a cast or ball, and stains hand.	0.0	Soil dark, feels smooth, and will form a ball; when squeezed, it ribbons out between fingers and leaves wet outline on hand.	0.0	Soil dark, may feel sticky, stains hand; ribbons easily when squeezed and forms a good ball.	0.0
Soil dark, stains hand slightly; forms a weak ball when squeezed.	0.3	Soil dark, feels slick, stains hand; works easily and forms ball or cast.	0.5	Soil dark, feels slick, stains hand; ribbons easily and forms a good ball.	0.7
Soil forms a fragile cast when squeezed.	0.6	Soil crumbly but may form a weak cast when squeezed.	1.0	Soil crumbly but pliable; forms cast or ball, will ribbon; stains hand slightly.	1.4
Soil dry, loose, crumbly.	1.0	Soil crumbly, powdery; barely keeps shape when squeezed.	1.5	Soil hard, firm, cracked; too stiff to work or ribbon.	

This guideline does not work, however, for fields treated with a growth regulator such as mepiquat chloride (Pix), because the stems of treated plants are normally reddish to within 2 inches (5 cm) of the tips. Stem color is not reliable in hot desert areas, where it is better to monitor wilting. Plants normally wilt to some extent on hot afternoons, especially in coarse soils, but wilting at noon generally means that water is needed soon. Wilting during morning hours means irrigation is overdue and/or that root systems have been injured by root-knot nematodes or other root problems. The "flower garden" effect that occurs when numerous flowers are visible at the tops of the plants also indicates that plants are stressed.

A crop's water needs usually vary from one part of a field to another, especially if the field includes different soils. Plants in a sandy streak or where root development has been restricted will show stress sooner than will the rest of the crop. Watch these weak areas to gain advance notice of when irrigation is needed for the rest of the field. However, schedule irrigations according to the need shown by most of the crop, not according to the condition of a relatively small number of plants.

A more precise measure of water stress in plants is the *leaf water potential* (LWP). LWP is usually expressed as a negative value in units called *bars*; the larger the negative number, the greater the water stress in the plant. Just after irrigation, a typical LWP reading is −11 bars; growth in cotton stops when LWP reaches about −24 bars.

Devices called pressure bombs use the difference between atmospheric pressure and the pressure inside a leaf as a measure of LWP. Directions for using them are provided in *Water Management for Cotton* (see References). Pressure bombs have been used successfully for scheduling irrigations on Acala SJ-2 cotton in the San Joaquin Valley. The optimum time for first postplant irrigation there is generally when the LWP reaches −16 bars. For midseason irrigations, the reading should drop to −18 to −20 bars. Pressure bombs are less reliable in the desert valleys, apparently because greater variability in cloud cover, humidity, and temperature cause wide fluctuations in the atmosphere's evaporative power. If you use the pressure bomb to schedule irrigation, test various LWP levels in a limited area for a season or two before applying the system to an entire field.

Another technique for determining water status of plants is to measure the canopy temperature with a device sensitive to thermal infrared (IR) energy. The difference between the canopy temperature and the air temperature can be used in the same way as LWP when combined with a measure of the atmospheric vapor deficit, or drying power of the air. One advantage of the IR technique is that you can take a large number of readings quickly. However, the technique provides little warning of the crop's need for water and it is not reliable when atmospheric conditions are variable. More study is needed before a specific procedure can be recommended for using IR to schedule irrigations in western cotton.

Instruments that measure the moisture content of the soil, such as tensiometers, gypsum blocks, and neutron probes, can help determine the need for irrigation, although the information they provide is not as direct as that obtained from the plant. To obtain reliable readings, you must install these instruments in soil that is typical of the whole field. Readings from an instrument placed in a small sandy streak, for example, would be misleading. In using these instruments, follow the recommendations of your supplier and your Cooperative Extension agent.

A convenient way to monitor soil moisture indirectly is to use a *water budget* to estimate how much water the crop uses from day to day under prevailing weather conditions. After soil has drained to field capacity, further loss of soil water occurs mainly through evaporation from the

FIELD NO. _122 A_

ALLOWABLE DEPLETION _2 inches_

DATE		ET IN INCHES PER DAY		CUMULATIVE ET
5/2		irrigation		
5/3		0.12		0.12
5/4		0.15		0.27
5/5		0.15		0.42
5/6		0.17		0.59
5/7		0.19		0.78
5/8		0.21		0.99
5/9		0.20		1.19
5/10		0.25		1.44
5/11		0.25		1.69
5/12		0.29		1.98
5/13		irrigation		

Figure 17. If you have a source of evapotranspiration (ET) data, you can use a water budget to estimate the interval between irrigations. Start on the day after irrigation to keep a running total of daily ET figures. When the total approaches the allowable depletion—2 inches in this example—check soil moisture in the field to make a final decision on when to irrigate.

soil surface or by transpiration from leaves. The combination of evaporation and transpiration is called *evapotranspiration* (ET). If you know how much available water is in the crop's rooting depth at field capacity, and if you know how much water is lost through ET each day, you can then estimate the amount of available water remaining at any time by adding up the daily ET values (Figure 17). Newspapers and radio stations in some areas report ET figures for major crops, including cotton.

Managing Salinity. Prevention of harmful salt buildup is an important part of irrigation. As part of routine soil analysis, have a laboratory determine the concentration of total salts and the proportion of sodium in the soil. Test irrigation water for total salts and sodium every 2 or 3 years, especially if it comes from wells where the water table is dropping. Test each well separately, as salt concentrations may differ greatly, even among wells in the same area. If the salt concentration in the soil is high, plan to leach the salt below the root zone with extra water, either during preirrigation or at the first irrigation.

Irrigation water carries salts as it soaks through soil; where the "front" of water stops, a high salt concentration may remain. If the front stops directly in the seed row, the resulting salt concentration can retard germination and injure seedlings. During early irrigations, make sure the front stops before reaching the seed row or soaks past it. Where drip irrigation emitters are placed in the seed row, the water carries salts away from the plants as it soaks outward. Water from emitters placed in furrows, however, moves salts toward the seed row. In either case, extra water is needed occasionally to leach salts from the root zone.

Fertilization. To produce a profitable cotton crop, make sure adequate nutrients are present in the soil. When nutrients are lacking, cotton competes poorly with weeds and is less able to outgrow disease and insect damage. Severe deficiencies may prevent growth entirely. Nitrogen fertilizer is needed every season in nearly all fields, and extra potassium, phosphorus, or zinc can improve yields in some soils. Western cotton soils usually contain adequate amounts of other nutrients.

Nitrogen. You must replenish nitrogen each season because the amount removed by the crop is large compared with the amount available in soil. Also, nitrogen is lost due to leaching and may be removed from soil by bacteria. The amount of nitrogen needed varies according to the crop rotation sequence, soil type, and yield potential. Rates as low as 75 pounds per acre are adequate in New Mexico and in some parts of the San Joaquin Valley. As much as 300 pounds per acre may be needed for maximum yield in low desert soils where cotton has a second fruiting cycle. Lower rates may be adequate where relatively large amounts have remained in the soil from the previous season. Where a stand of alfalfa equivalent to a cutting of hay has been turned under before planting, there may be adequate soil nitrogen to produce a cotton crop without extra fertilizer.

Fertilizers supply nitrogen (N) in the form of ammonium (NH_4^+), nitrate (NO_3^-), or both. Urea, another form of nitrogen found in some fertilizers, is converted to ammonium in the soil. Cotton plants can use both ammonium and nitrate, but they absorb nitrate more quickly because it remains in solution in soil water, while ammonium adheres to clay particles. Because of its greater solubility, nitrate also leaches from soil more rapidly than ammonium.

Soil bacteria convert ammonium to nitrate in two steps. One group of bacterial species first changes ammonium to nitrite (NO_2^-). A second group changes nitrite to nitrate. In oxygen-poor conditions found in waterlogged soil, certain bacteria convert nitrate to gases such as nitrogen (N_2) and nitrous oxide (N_2O); these escape into the atmosphere, resulting in substantial depletion of soil nitrogen.

Efficient use of fertilizer requires that adequate nitrogen be available when it is most needed and that excessive nitrogen not remain at the end of the season. Cotton uses little nitrogen during seedling growth; most of the nitrogen needed is used during fruiting (Table 3). Excessive nitrogen or nitrogen applied too late in the season can delay maturity and may stimulate growth of unnecessary stems and leaves that end up as trash at harvest. Where nitrogen is applied too late, extra applications of defoliant may be necessary. Cotton that is lush due to excessive nitrogen or water favors lygus bugs, bollworms, tobacco budworm, pink bollworm, and boll rots.

Table 3. Use of Nutrients by Cotton during Four Stages of Growth. Data from studies of Acala cotton in the San Joaquin Valley.

Stage of Growth	PERCENT OF SEASONAL REQUIREMENT		
	Nitrogen	Phosphorus	Potassium
germination to first square	10%	7%	7%
first square to first boll	30	31	23
first boll to open boll stage	40	35	53
open boll stage to maturity	20	27	17

Preplant soil tests are valuable for determining early season nitrogen requirements. They do not measure nitrogen levels precisely, because much soil nitrogen is in a form not readily available to plants. However, tests can determine whether relatively large or small amounts of

nitrogen remain from the previous season, and they can assist in scheduling early season fertilizer applications. Table 4 shows the application schedule recommended in Arizona, based on results of preplant soil tests. For recommendations in other areas, check with farm advisors or county agents.

Table 4. Arizona Recommendations for Nitrogen Applications Based on Results of Preplant Soil Tests.

Nitrate Nitrogen ($NO_3 - N$) in Preplant Soil Test (Parts per Million)	Recommendation
0 to 10	apply nitrogen before planting or soon after seedlings emerge
10 to 20	apply nitrogen by first square
20 to 30	apply nitrogen by first bloom
above 30	use petiole analysis to determine need for nitrogen

Analysis of petiole tissue is the best way to determine the crop's nitrogen needs during the growing season. Take petiole samples for laboratory testing at least three times each season; contact the lab in advance for special instructions. Take samples when soil moisture is moderate; nutrient availability is affected by moisture, so test results may be misleading if samples are taken immediately after irrigation or when soil is too dry.

Analysis generally requires 30 to 50 petioles from each field or each area that differs in soil type or crop growth. Pick petioles one at a time from separate plants as you walk through the field. From each plant, pick the petiole of the youngest fully expanded leaf on the main stem; this is usually leaf number 4 or 5 if you count the newest partly unfurled leaf as number 1 (Figure 5, page 12). If you are uncertain which of two leaves to pick, take the older one. Discard the leaf blades unless the lab needs them for special tests; put the petioles in a small, clean paper bag, label them, and deliver them to the lab immediately.

Tables 5 and 6 show the nitrogen levels, expressed as parts per million of nitrate nitrogen, considered deficient and adequate for cotton in the San Joaquin Valley and in Arizona. Similar information has not been developed for New Mexico, but nitrogen levels should be approximately the same as in the San Joaquin Valley.

Tissue tests early in the season can help in scheduling nitrogen applications during the current year; later tests are useful for determining fertilizer needs the following year. It is usually best to see the results of two subsequent sample dates before changing application rates or schedules. If nitrogen is needed to increase plant nitrate levels, remember that ammonium must be converted to

Table 5. Nutrient Levels in Cotton Petiole Samples in the San Joaquin Valley. When nutrient levels are below those in Low columns, a higher fertilizer rate will usually improve yield. Values in the High columns represent the upper end of the normal range for each nutrient.

Time of Sampling	NITRATE NITROGEN ($NO_3 - N$) parts per million		PHOSPHATE PHOSPHORUS ($PO_4 - P$) parts per million		POTASSIUM (K) percent	
	Low	High	Low	High	Low	High
first bloom—approximately July 1	10,000	18,000	1,500	2,000	4.0	5.5
peak bloom—about 30 days after first bloom or about July 31	3,000	7,000	1,200	1,500	3.0	4.0
late bloom—about 60 days after first bloom or approximately August 30	1,500	3,500	1,000	1,200	1.5	2.5
late season—approximately 70 days after first bloom or about September 10	Should fall below 2,000		800	1,000	1.0	2.0

Table 6. Petiole Nitrate Levels Considered Adequate and Inadequate for Cotton under Arizona Conditions.

Time of Sampling	NITRATE NITROGEN ($NO_3 - N$) LEVELS IN PETIOLES Inadequate if Below	Adequate
early squaring	16,000 ppm	18 to 22,000 ppm
first bloom	12,000	13 to 15,000
first green boll (15 days old)	6,000	7 to 12,000
first open boll	4,000	5,000
end of season	should drop below 2,000 ppm	

nitrate and must be leached into the root zone with irrigation before it becomes available to the crop. Nitrogen supplied as nitrate is available more quickly. If the nitrate level is above about 2500 ppm at the last sample date, the excess nitrogen may interfere with defoliation and will be wasted if leached by winter rains. To avoid this the following season, reduce the nitrogen rate or schedule earlier applications.

In most cases, enough soil nitrogen is present at planting to support seedling establishment; preplant application is seldom necessary. A more efficient way to supply nitrogen is to apply it as a side-dressing after seedlings are established. The fertilizer is then less exposed to leaching during early irrigation and a greater percentage is likely to remain during fruiting. Two or more side-dressings may be needed, depending on soil type and length of the season; more frequent but smaller applications are needed in sandy soils, and extra nitrogen is usually needed in long-season crops.

Although side-dressing is generally best, it is also possible to apply a large portion of the season's nitrogen before planting. Preplant application is more successful in fine-textured soils than it is in coarse-textured soils, where nitrogen is more subject to leaching. To apply ammonium fertilizer to alkaline soil as a broadcast treatment, work the fertilizer into the soil with discs or similar tools; otherwise, there can be a substantial loss of nitrogen as ammonia gas.

Another way to apply nitrogen is through irrigation. In furrow-irrigated fields, there is no way to assure uniform distribution of fertilizer applied in the water; the tail ends of runs may receive excessive amounts, while other parts of the field may remain deficient. Drip irrigation, however, provides an excellent way to apply nitrogen uniformly.

Other Nutrients. Aside from nitrogen, phosphorus is the nutrient most often needed for high cotton yield in western soils. Preplant soil tests can determine whether phosphorus is needed in the current season, and late-season tissue samples are good guides to the need for phosphorus the following year. About 40 to 60 pounds of phosphorus pentoxide (P_2O_5)—the equivalent of 88 to 133 pounds of treble superphosphate—is a typical rate for preplant applications.

Potassium deficiency occurs in certain San Joaquin Valley soils, apparently due partly to the interaction of Acala cottons with a soilborne fungus (page 110). In these situations, very large amounts of potassium may be needed to eliminate the deficiency. Most soils elsewhere in the Western Region contain adequate potassium.

Zinc is needed most often in alkaline soils high in sodium or calcium carbonate (lime). Only occasionally are other soils deficient. A single preplant application of 40 to 50 pounds of zinc sulfate per acre usually provides adequate zinc for 2 to 3 years.

Fertilization with phosphorus, zinc, or potassium is most efficient if they are added before planting or as side-dressings before squaring. Adding them as side-dressings later may not allow time for them to affect plant growth fully during the current season. Foliar sprays can correct deficiencies of zinc and other nutrients that are needed only in small amounts.

In the San Joaquin Valley, a range of soil test values has been established which indicates whether a yield increase can be expected by adding phosphorus, potassium, or zinc in fertilizer (Table 7); for phosphorus and potassium, there is also a set of plant tissue values considered deficient or adequate (Table 5). Similar values have not been established for other areas, but farm advisors and consultants can help determine the need for these nutrients in local soils. Phosphorus, potassium, and zinc are relatively stable in soil, so you need not test for them every season. After initial analysis, retesting every 2 to 5 years is adequate. Always test soil following alfalfa and other perennial crops.

Table 7. Soil Nutrient Levels for Cotton in the San Joaquin Valley. When nutrient levels are below those in the first column, addition of fertilizer usually improves yield. At levels above those in the second column, fertilizer is not likely to produce a measurable response. Nutrients are measured as follows: phosphorus as $NaHCO_3 - P$; potassium as exchangeable K (ammonium acetate test); and zinc as DTPA-extractable Zn.

| | YIELD RESPONSE | |
| | LIKELY | NOT LIKELY |
Nutrient	parts per million in extract	
phosphorus	5	8
potassium:		
sandy soils	40	60
loam soils	40	80
clay soils	60	100
zinc	0.4	0.7

Plowdown. A prompt, thorough plowdown is essential for all cotton crops. Cooperate with local authorities by observing plowdown regulations, which are indispensable to preventing the establishment and spread of such major insect pests as pink bollworm and boll weevil.

Plowdown involves the shredding and burial of all cotton stalks, unharvested bolls, and other crop debris immediately after harvest. Shredding destroys some insect pests directly and makes it possible to bury debris uniformly in discing. It also promotes rapid decay of debris, thereby making it easier to shape seedbeds, to cultivate, and to apply herbicide uniformly. After shredding, cross-disc or plow the field to turn under crop debris and surface soil to a depth of 6 inches (15 cm). Overwintering pink bollworms and boll weevils cannot emerge in spring if they are buried that deeply. Destruction of overwintering in-

dividuals is essential in keeping populations down to levels that can be managed economically. Even if you do not plan to replant cotton in the same field, thoroughly plow down the crop; abandoned crops are a potential source of insect pests for a large surrounding area.

When plowing down a cotton crop, destroy stands of weeds in fencerows and ditchbanks so that they cannot serve as a source of seeds or vegetative spread into next season's crop. Eliminate volunteer cotton along with other weeds.

Pesticides

Properly used, pesticides can provide economical protection from pests that otherwise would cause significant loss. In many situations, they are the only feasible means of control. Careless or excessive use of pesticides, however, can result in poor control, crop damage, higher expenses, and hazards to health and the environment. In an IPM program, pesticides are used only when field monitoring indicates they are needed to prevent losses.

In choosing a pesticide, consider not only its effect on the target pest, but also the effects it may have on other pests, natural enemies, and crop plants. Effects may vary according to formulation, rate, and application method. Consult the latest Cooperative Extension recommendations before making a choice, and keep in touch with farm advisors or county agents for current information on new materials and methods.

Make sure each pesticide treatment is aimed at the appropriate stage of the target pest and is timed to coincide properly with crop growth, weather conditions, and cultural operations. You can then get the level of control you need with the fewest treatments, thereby reducing costs and potential hazards. Also, the chance of such side effects as pest resurgence, secondary pest outbreaks, and pesticide resistance will be lessened.

Always READ THE LABEL before using any pesticide. Follow directions carefully and observe suggested safety precautions.

Pesticide Resistance. Some cotton pests, mostly insects and mites, have developed resistance to certain pesticides; they are able to survive applications that once killed most individuals of the same species (Table 8). At first, it may be possible to control resistant pests by increasing rates, but higher rates and more frequent exposure may increase the chance that resistant individuals will increase as a proportion of the total population (Figure 18). When that happens, the pesticide involved will no longer be economically feasible. You may then need to switch to a new material. The search for new pesticides is complicated by the fact that species resistant to one material often develop *cross-resistance* to others, even to materials not in the same chemical class. Resistance to new pesticides may then develop much more quickly than it did with the original one. For example, insect pests that have developed resistance to chlorinated hydrocarbons such as DDT are more likely to become resistant to synthetic pyrethroids.

Pesticide resistance can develop in a pest not intended as the target of applications. Whiteflies and the cotton leaf-

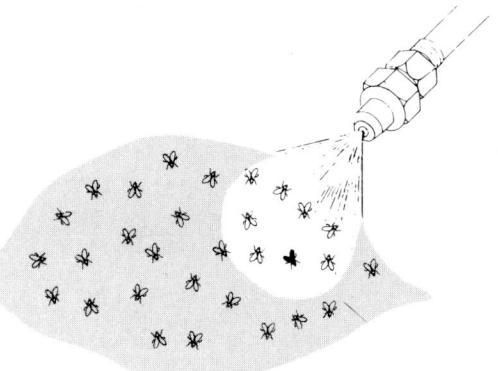

Some individuals in a pest population have genetic traits that allow them to survive a pesticide application.

A proportion of the survivors' offspring inherit the resistance traits. At the next spraying these resistant individuals will survive.

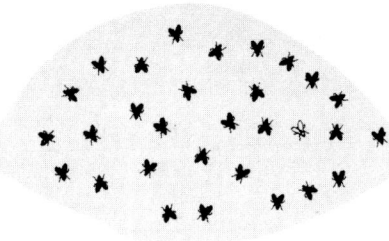

If pesticides are applied frequently, the pest population will soon consist mostly of resistant individuals.

 susceptible individual

resistant individual

Figure 18. Pest populations develop resistance to pesticides through genetic selection.

Table 8. Pesticide Resistance Reported in Insects and Mites of Cotton in the United States. Resistance in a particular species does not necessarily occur uniformly in all areas.

Pest Species	Pesticide
Caterpillars	
bollworm	chlorinated hydrocarbons methyl parathion carbaryl methomyl
tobacco budworm	chlorinated hydrocarbons organophosphates carbaryl methomyl
beet armyworm	chlorinated hydrocarbons methyl parathion
cabbage looper	chlorinated hydrocarbons organophosphates
cotton leafperforator	chlorinated hydrocarbons organophosphates carbamates
pink bollworm	DDT
saltmarsh caterpillar	chlorinated hydrocarbons
Sucking Insects	
bandedwing whitefly	methyl parathion
cotton aphid	benzene hexachloride
cotton fleahopper	chlorinated hydrocarbons organophosphates
lygus bugs	DDT malathion monocrotophos trichlorfon
southern garden leafhopper	DDT
stink bug, *Euschistus conspersus*	chlorinated hydrocarbons
Other Insects	
boll weevil	chlorinated hydrocarbons
thrips	chlorinated hydrocarbons
Spider Mites	
carmine, pacific, strawberry, and twospotted mites	organophosphates
pacific and twospotted mites	dicofol

perforator, for example, have developed broad resistance because of frequent insecticide applications aimed at other pests.

Pest Resurgence and Secondary Outbreaks. Pesticides that kill or disturb natural enemies may cause pest resurgence or outbreaks of secondary pests. This often happens with use of insecticides.

Pest resurgence occurs when a pesticide destroys natural enemies of the target pest. Because the natural enemies depend on the pest for food, they take longer to build up to their former numbers. On the other hand, pests that survive treatment or that move into the field later can breed without the restraint of natural enemies, sometimes increasing to greater numbers than existed before treatment. Resurgence of bollworms and tobacco budworms is common in western cotton.

Pesticide applications may also cause damaging increases in pests that were not the target. Such increases, called secondary outbreaks, usually result from destruction of natural enemies that controlled the secondary pest before the application (Figure 19). Secondary outbreaks of spider mites are common in the San Joaquin Valley following treatments for lygus bugs. Other pests that often increase following insecticide treatments include bollworms, tobacco budworm, beet armyworm, cabbage looper, cotton leafperforator, and whiteflies. Some insecticides promote outbreaks of mites by stimulating their reproduction directly or by causing changes in plant physiology that favor them, as well as by destroying natural enemies.

Crop Injury. Crop injury (phytotoxicity) due to pesticides can result from an improper application method, poor timing of application, excessive rates, drift, and residues in soil or water. Most common problems are due to herbicides. Some preplant herbicides may injure seedlings if the seed is planted too shallowly in treated soil (Figure 50, page 120) or if emergence is delayed by cool weather; using a material or rate not suited to local conditions can reduce stands. Herbicides applied after cotton emergence may injure the crop if they are applied at the wrong stage of growth or without the proper equipment needed to protect the crop.

If you plan to follow cotton with a rotation crop, check labels to make sure the herbicides you use in cotton will not leave a soil residue harmful to the next crop. Before planting cotton, be sure you know which herbicides were used in the field the previous season. Don't use irrigation water that may contain harmful residue in runoff.

Under certain circumstances, some insecticides can also injure cotton. Methomyl (Lannate/Nudrin) may cause a reddening of leaves that resembles spider mite injury, especially if it is applied more than twice. Multiple

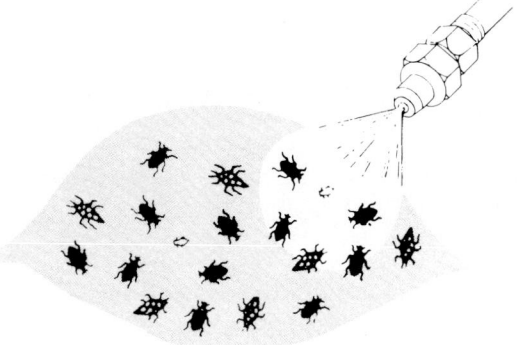

A pesticide applied to control pest A also kills natural enemies that are controlling pest B.

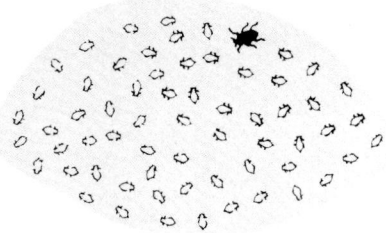

Released from the control exerted by natural enemies, pest B builds up to economically damaging levels

   pest A

   pest B

natural enemy

Figure 19. Secondary outbreaks of insects and mites are often caused by destruction of natural enemies.

applications of methyl parathion or carbaryl during or before peak bloom may delay fruiting and have reduced yields in some cases. Preplant application of disulfoton (Di-Syston), phorate (Thimet), chlorpyrifos (Lorsban), or aldicarb (Temik) may contribute to reducing stands when seedling growth is slowed by cool weather. The injury is due to the interaction of chemical effects with low temperature and seedling disease pathogens.

Effects on Other Crops. Adverse effects are not always limited to the fields where pesticides are applied. Resistant pest populations with a wide host range, such as whiteflies or the tobacco budworm, may affect many crops, not just the one where the resistance developed. Pesticides may also destroy natural enemies that would otherwise colonize and protect neighboring crops. Drift, particularly from aerial herbicide applications, can endanger nearby crops.

Hazards to Human Health. Some pesticides used in cotton are hazardous to humans. Most at risk is the applicator; field personnel, irrigators, and others who enter the field may also be exposed. Pesticides that drift onto roadways, yards, or other areas outside the field may also cause health problems. Read and follow the safety precautions on pesticide labels and consult the Cooperative Extension Service for guidelines on how to use pesticides safely. Several safety publications are listed in the References.

Before applying a pesticide, call a local doctor or hospital to confirm that emergency medical care is available. Make sure application equipment is in good condition and has the appropriate safety features for the material you are using. Always wear appropriate gloves, mask, and other protective gear when handling or applying pesticides.

Hazards to Bees. Insecticides applied to cotton have caused major losses to beekeepers in some areas. Although bees are not important in pollinating cotton, they forage for nectar when cotton is in bloom and they may also be exposed to insecticides through drift. To reduce hazards to bees, treat at night or before dawn, when bees are not active; use materials that are relatively safe for bees; and avoid letting insecticides drift to flowering crops or weeds or to water sources visited by bees. Cooperate with local authorities in notifying beekeepers of insecticide applications that may affect bees.

The degree of an insecticide's hazard to bees depends not only on its toxicity, but also on rates, method and timing of application, and formulation. Under field conditions, many organophosphates are highly hazardous to honeybees. These include parathion, methyl parathion, acephate (Orthene), dimethoate (Cygon), EPN, diazinon, monocrotophos (Azodrin), azinphos-methyl (Guthion), naled (Dibrom), chlorpyrifos (Dursban), phosmet (Imidan), and phosphamidon (Dimecron). Other insecticides especially hazardous to bees include carbaryl (Sevin) and oxamyl (Vydate), both carbamates. Synthetic pyrethroids used in cotton are moderately hazardous if applied when bees are present, but are relatively safe if applied at night. The microbial insecticides, *Bacillus thuringiensis* (Dipel/Thuricide) and *Heliothis* nuclear polyhedrosis virus (Elcar), are not toxic to bees.

Dusts and wettable powders are generally more hazardous than emulsifiable or soluble formulations of the same material. Also, sprays of large droplet size are more hazardous than are those with small droplets. For more information, see the publication, *Reducing Pesticide Hazards to Honey Bees*, listed in the References.

Hazards to Wildlife. Pesticides applied to cotton can contaminate water, and fish, birds, or other animals may be killed if the contaminated water drains into streams or other bodies of water. Don't apply pesticides when irrigation water may run off into streams, and don't let pesticides drift onto water. Keep abreast of regulations designed to protect wildlife from pesticide contamination.

Vertebrates

Vertebrate pests injure western cotton only occasionally. Jackrabbits, *Lepus* spp., are the most common. Although cotton is not a favored food, jackrabbits will feed on it when other food is limited. Populations peak every few years, and damage is most likely during population peaks and in drought years when natural forage is scarce.

Jackrabbits will destroy cotton seedlings, sometimes leaving a swath of bare ground through a field of seedlings. Losses are sometimes extensive in isolated fields, usually those surrounded by large areas of open land. Although jackrabbits may find shelter in cotton fields later in the season, they cause little or no damage to mature plants, as they eat only the petioles of the lower leaves. Where they have fed on older plants, you may find discarded leaf blades on the ground.

Ground squirrels, *Spermophilus* spp., occasionally feed on young cotton plants, causing the same injury as jackrabbits. They may also pull mature bolls from the plants and carry them to their burrows to remove the seeds. Injury is usually limited to the edges of a field along roadsides or ditches where the burrows are located.

To prevent damage by jackrabbits and ground squirrels, locate their populations well before planting. Poison baits are the most effective controls, but baiting seldom prevents damage unless it begins well before a susceptible crop is present. Control of jackrabbits is usually needed only in years with peak populations, but control of ground squirrels may require two or more seasons to reduce populations significantly.

Other vertebrate problems in western cotton are very limited. Gophers can damage drip irrigation systems by chewing through drip lines, so it is a good idea to control them before installing a drip system. Mule deer sometimes eat leaves of cotton in plantings near Arizona's desert mountain ranges. Mice and other rodents sometimes live in weedy cotton fields; they cause little or no damage, but in desert areas they may attract rattlesnakes, which are a potential hazard to persons working in the field.

Figure 20. Ground squirrels may enter fields from uncultivated areas. They feed on young plants and bolls near the edge of the field.

Vertebrate damage is potentially greater in glandless cotton because glandless varieties lack the gossypol which deters feeding by many vertebrates. For more information on the vertebrate pests, consult sources listed in the References.

SEQUENTIAL SAMPLE CARDS

SAMPLE CARD A

threshold = 50% infested

sample no.	don't treat	tally	treat
1	—	0	—
2	—	1	—
3	—	2	—
4	—	3	—
5	—	3	—
6	3	4	7
7	4	5	7
8	4	5	8
9	4	6	9
10	5	6	9
11	5	7	10
12	6	8	10
13	6	9	11
14	7	10	11
15	7	10	12
16	8	11	12
17	8	12	13
18	9	13	13
19	9	___	14
20	9	___	15

Treat

SAMPLE CARD B

threshold = 50% infested

sample no.	don't treat	tally	treat
1	—	1	—
2	—	1	—
3	—	2	—
4	—	2	—
5	—	3	—
6	3	4	7
7	4	5	7
8	4	5	8
9	4	5	9
10	5	6	9
11	5	6	10
12	6	6	10
13	6	___	11
14	7	___	11
15	7	___	12
16	8	___	12
17	8	___	13
18	9	___	13
19	9	___	14
20	9	___	15

Don't Treat

SAMPLE CARD C

threshold = 50% infested

sample no.	don't treat	tally	treat
1	—	1	—
2	—	2	—
3	—	2	—
4	—	3	—
5	—	4	—
6	3	5	7
7	4	5	7
8	4	5	8
9	4	6	9
10	5	7	9
11	5	7	10
12	6	8	10
13	6	8	11
14	7	8	11
15	7	9	12
16	8	10	12
17	8	10	13
18	9	11	13
19	9	11	14
20	9	11	15

Recheck

Figure 21. To use a sequential sampling card, take a series of samples and keep a running tally of results in the center column. The kind of sample depends on the pest you are monitoring. As shown by dashes at the tops of the "treat" and "don't treat" columns, you must take a certain minimum number of samples before reaching a treatment decision; in this case the minimum is six. Beyond the minimum, you can reach a treatment decision whenever the number in your tally matches the corresponding number in one of the boundary columns. If the tally reaches the number in the "treat" column (Sample Card A), the pest population is above the treatment threshold. If the tally matches the "don't treat" column (Sample Card B), the population is below the threshold. Continue sampling as long as the tally remains between the boundary columns. If you reach the bottom of the card and the tally is still between the boundaries (Sample Card C), the population is too close to the threshold for you to make a reliable decision; return and sample again in 1 to 3 days.

Insects and Mites

Several hundred species of insects and mites are found in western cotton fields, but only a handful are destructive. Many are predators or parasites, some of them important natural enemies of the pests, and others visit cotton fields only for pollen or shelter. The most damaging pests are those that attack squares and bolls: the bollworm, tobacco budworm, pink bollworm, boll weevil, and lygus bugs. Leaf-feeding species are generally less important, although spider mites, the cotton leafperforator, and certain others can reduce yield if they destroy too much foliage. Seedling pests, such as cutworms, usually have only local impact, but dense, spotty infestations can reduce stands and can make it necessary to replant parts of a field. Whiteflies and aphids seldom affect yield but they can reduce the grade of lint by contaminating it with honeydew.

Because cotton is grown as an annual crop, most insects cannot survive all year in cotton fields, but they move in each season from other crops or from weeds. Alfalfa harbors numerous species all year; spider mites, lygus bugs, and beet armyworms, for example, move into cotton from alfalfa. Most major predators, including big-eyed bugs, minute pirate bugs, and damsel bugs, also come from alfalfa, as do many parasites that attack caterpillars. Safflower, sugarbeets, and vegetables are also potential sources of cotton insects. Weeds and native vegetation are important sources of lygus bugs, beet armyworm, and leafhoppers.

Monitoring

Most guidelines for using insecticides are based on sampling, which involves picking or examining specific plant parts or, in the case of lygus bugs, collecting insects with a sweep net. Pheromone traps and degree-day models help in monitoring population cycles, but they cannot be used directly for scheduling treatments.

In most areas, there are seldom more than two major pests present at once, so you can concentrate monitoring on these species. Where insecticides have been applied, however, you must also be alert for secondary outbreaks of other pests. Watch also for pests that occur only occasionally or only in limited areas such as at the edges of fields.

Sampling

Two kinds of sampling are used for monitoring insects and mites: standard sampling and sequential sampling. Most people are more familiar with *standard sampling*, which involves taking a fixed number of samples at each visit to the field. In monitoring for bollworms, for example, you might check 100 terminals each time and count the larvae present. In *sequential sampling*, you take samples only until the results show that the pest population is either above or below the threshold level.

The advantage of sequential sampling is that it concentrates sampling time where it is most needed—in fields where the pest population is close to the treatment threshold. When a population is either very high or very low, it takes only a few samples to show that it is above or below a threshold. For a population near the threshold, on the other hand, more samples are needed. For sequential sampling, use a card such as those in Figure 21. The "No Treat" and "Treat" columns are boundaries set by the treatment threshold for each pest and by the error rate—the chance of making a mistake.

In either kind of sampling, there are often borderline cases in which you cannot be certain whether the population is above the threshold or not. In these cases, it is usually best to return in 2 or 3 days to take another set of samples.

Taking Samples Randomly. Look hard enough for pests, and you will probably find a few, but this does not mean the population will cause economic loss. Useful evaluation of pest populations requires that you choose sample plants *randomly* and avoid singling out plants more likely than others to be infested. Most people have an unconscious tendency to examine tall plants when checking for pests. Because tall plants are more likely to be infested with certain pests, this tendency is an important source of error. Another common mistake is to choose only obviously damaged plants.

To ensure that sample plants are chosen randomly, first walk 50 paces into the field, then stop and find the plant closest to your foot. Count down the row to the tenth plant from the closest one; this is the plant to sample. To check several plants at each stop, use the same procedure to find the first one, then check the required number of adjacent plants. Step off a fixed number of paces (20 or 30) between each sample site, and count down the row from the closest plant each time to find the sample plant. As you move, zigzag to cover as much of the field as possible. Be sure to include some samples from the head, tail, and middle of irrigation runs.

Sample Area. Any decision based on sampling applies only to the field sampled and only to that part where conditions are the same as in the area sampled. In a field with areas that differ in plant growth because of variable soil type, drainage, or other factors, treat each area separately. For example, if one section with poor soil has smaller plants than elsewhere, take a separate set of samples there. If a pest population is above the treatment threshold only in one area, apply pesticide only to that area. There is no rule on how large an area can be evaluated with a certain number of samples, but the more uniform the field is, the larger the area can be.

The Edge Effect. Many infestations begin at the edges of fields where pests have moved in from adjacent crops or weeds. Some pests remain concentrated at the edges throughout the season. Because of this "edge effect," you should not take samples at the edge unless you plan to treat it separately. If you are monitoring the field as a whole, always walk in at least 50 paces before you begin sampling; otherwise, samples taken close to the edge may bias the result. This does not mean, however, that you should ignore the edges; it is a good practice to check them, especially those bordering alfalfa or weedy areas, at each visit. Watch for infestations of mites, lygus bugs, and other pests that may later spread. Some pests, such as yellowstriped armyworms and stink bugs, are often limited to field edges, where they can be controlled with strip treatments.

Sampling Error. There are two ways to make a mistake in sampling, and both can be expensive. If you decide incorrectly that an infestation is below the treatment threshold, losses due to pest damage may result. On the other hand, if you decide that it is above the threshold when it is not, costs may include the expense of an unnecessary treatment as well as possible deferred costs due to pest resurgence, secondary outbreaks, and development of pesticide resistance.

You can reduce sampling error by taking more samples, but the cost of sampling may then exceed the cost of making a mistake. This is why the error rate allowed in sampling plans for pest management—although it is not always specified—is usually 20% or more.

A 20% error rate means there is one chance in five of deciding that the pest population is below a threshold when it is actually above it, or vice versa. However, the chance of making a *serious* mistake is much lower. Most errors occur when the population is close to the threshold, when the impact of a wrong decision is small. The chance of making a mistake when the population is well above or below the threshold is much lower.

Seasonal Monitoring Schedules

Each western cotton-growing area has its own sequence of pest problems and requires a different schedule of monitoring activities. These are summarized in Figure 13 (page 20).

San Joaquin Valley. In the San Joaquin Valley, no single pest species occurs in damaging numbers every year, but spider mites and lygus bugs occur regularly enough that careful monitoring is always needed. Early treatments for lygus bugs often result in outbreaks of spider mites, and late lygus treatments may contribute to bollworm infestations. Widespread outbreaks of beet armyworms and loopers occur occasionally, but damaging populations appear only locally in most seasons, usually following treatments for other pests. The pink bollworm has appeared in the San Joaquin Valley in small numbers, but the potential for its permanent establishment is uncertain. Because of the San Joaquin Valley's shorter growing season, the pink bollworm probably would not have the impact it has had in the desert valleys. Observing local plowdown and planting regulations is a worthwhile precaution, however, to limit the chance that damaging populations could develop.

The following monitoring practices are recommended for the San Joaquin Valley:

• Start before planting to watch adjacent crops and weeds for pests that could injure cotton. Pay special attention to alfalfa, where infestations of lygus bugs, spider mites, and yellow-striped armyworms often begin. Use a sweep net to find lygus bugs on weeds, especially on mustard family plants such as London rocket. Watch for larvae and egg masses of beet armyworm on pigweeds, nettleleaf goosefoot, and related weeds.

• As soon as seedlings emerge, walk the field once a week to look for cutworms and other seedling pests (page 86). If there are parts of a field where these problems have occurred before, check them more often.

• Start sampling for spider mites soon after crop emergence. Pay special attention to fields that have been treated with insecticides (page 73).

• As soon as squaring begins, use a net to look for lygus bugs on small plants. There is no treatment threshold during the first 2 weeks of squaring.

• From the third through the sixth weeks of squaring, take sweep net samples for lygus bugs once or twice a week. Count the squares on samples of plants weekly to determine the treatment threshold (page 64).

• After the sixth week of squaring, use weekly sweep net samples to check for lygus bugs, but drop the square counts.

• In the first week of August, start checking once a week for small bollworms on squares and young leaves at the tops of plants (page 44). Continue until most bolls have matured.

• Throughout the season, be alert for outbreaks of leaf-feeding caterpillars, whiteflies, aphids, and other occasional pests.

Desert Valleys. In the desert valleys of Arizona and southern California, the pink bollworm is a key pest that can cause unacceptable damage every year if not controlled. Insecticides used for pink bollworm often cause damaging outbreaks of such secondary pests as tobacco budworm, cotton leafperforator, and whiteflies. Reducing these outbreaks requires that insecticides be applied for pink bollworm only when sampling shows that the number of larvae in bolls has reached the treatment threshold. The boll weevil has also been very destructive in parts of western and central Arizona, although its long-term status there is still uncertain. Prompt shredding and plowdown and, when possible, early harvest are valuable in reducing overwintering populations of both pests.

Follow these steps for monitoring in the desert valleys:

• Beginning January 1, keep a record of degree-days so that you can estimate when overwintering pink bollworms will emerge.

• Check for pests in surrounding crops and weeds and monitor for seedling pests in the same way as suggested for the San Joaquin Valley.

• Set out pheromone traps for pink bollworm at or before first square (page 52).

• Starting in the third week of squaring, take weekly sweep net samples for lygus bugs (page 66). Watch for other pests and beneficials in the sweep net also.

• Start at first bloom to check plant terminals for small bollworms and tobacco budworms, and set out pheromone traps for these species (page 44). Continue sampling until most bolls have matured.

• As soon as 14-day-old bolls are present, take samples once or twice a week to check for pink bollworm larvae (page 45). Continue as long as susceptible bolls are present.

• From mid-to late season, watch for spider mites and leaf-feeding caterpillars, especially where insecticides have been used. If the cotton leafperforator is present, monitor it carefully; populations can increase very rapidly.

• In areas where the boll weevil is present, follow monitoring directions on page 60.

Central and Western New Mexico/Eastern Arizona.
The most frequent pests in central and western New Mexico are lygus bugs and bollworms. Because most New Mexico cotton fields are relatively small and because alfalfa is grown extensively, most fields have populations of natural enemies that keep other pests under control. Spider mites occur sporadically, most often following treatments for lygus bugs, and bollworms reach damaging levels in occasional years. Damaging populations of pink bollworm occur in occasional seasons following a series of mild winters.

These monitoring practices are recommended for New Mexico:

• Start accumulating degree-days on January 1 to monitor emergence of overwintering pink bollworms (page 52).

• Check for pests in surrounding crops and weeds and monitor for seedling pests as suggested for the San Joaquin Valley.

• At first square, start taking sweep net samples for lygus bugs twice a week, and continue through the squaring period (page 66).

• Set out pheromone traps for pink bollworm at first bloom (page 52).

• Start cracking bolls to check for pink bollworms as soon as 14-day-old bolls are present (page 52).

• Start checking the tops of plants for bollworms on July 1; continue until bolls begin to crack (page 44).

• Be alert all season for spider mites, beet armyworms, loopers, and other occasional pests, especially in fields where insecticides have been applied.

Growing conditions in southeastern Arizona are essentially the same as in central and western New Mexico. The pink bollworm is eastern Arizona's major cotton pest, but it has fewer generations there than in western Arizona because of the shorter season. Fewer insecticide treatments are needed, and secondary pests are much less damaging. The status of other pests in eastern Arizona is similar to that in central New Mexico. Traditionally, however, management recommendations have been organized state-by-state, so different sets of recommendations have been used on each side of the state border.

Eastern New Mexico.
Cultural methods and pest problems in eastern New Mexico are more similar to those in western Texas than to the rest of New Mexico. Lygus bugs and cotton fleahoppers are major pests, and bollworms occur regularly in eastern New Mexico, rather than occasionally as in the rest of the state. The tobacco budworm occurs in eastern New Mexico in small numbers. Pink bollworm appeared in damaging numbers in the late 1960s and early 1970s, but it has remained at low levels following establishment of a plowdown program.

Monitoring practices in eastern New Mexico are similar to those for central New Mexico, except that monitoring for pink bollworm is performed on an area-wide basis by Cooperative Extension personnel rather than by individual crop managers. Monitoring for cotton fleahopper is the same as for lygus bugs.

Eyes of bigeyed bug bulge out beyond the edge of the thorax. The tip segment of the antenna is slightly enlarged, and much of the body is covered with small pits. This species is *Geocoris punctipes*.

Eggs of bigeyed bugs are laid singly on leaves, often in mite colonies. They develop a red eyespot soon after they are laid.

A bigeyed bug nymph in the third instar. Like adults, nymphs have bulging eyes and short antennae with enlarged tips. The color is variable.

Minute pirate bug nymphs are often yellow or orange. This one is feeding on a spider mite.

Adult minute pirate bugs have an X-shaped black pattern on the back.

MINUTE PIRATE BUG

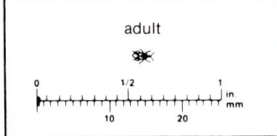

adult

Lacewing larvae use sickle-shaped, piercing mandibles to drain fluids from prey.

BIGEYED BUG	LACEWING
adult	larva

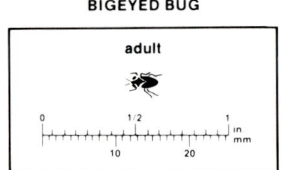 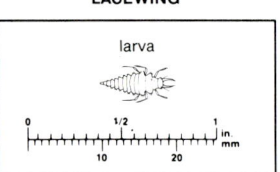

General Predators

General predators are those that feed on a wide variety of prey. Important general predators in western cotton are bigeyed bugs, minute pirate bugs, damsel bugs, lacewing larvae, and collops beetles. These species help limit populations of spider mites, caterpillars, aphids, and other pests. Assassin bugs and spiders are also general predators, but have less impact. Other important predators, such as lady beetles, syrphid fly larvae, and thrips, feed only on certain prey and are discussed elsewhere in this chapter.

Bigeyed Bugs   *Geocoris* spp.

Bigeyed bugs are among the most important natural enemies of cotton pests in the western U.S. They feed on the eggs and small larvae of bollworms, tobacco budworms, and other caterpillars; on eggs and nymphs of lygus bugs, other plant bugs, whiteflies, and leafhoppers; and on all stages of spider mites and aphids.

The two common bigeyed bugs, *Geocoris punctipes* and *G. pallens*, are light brown, tan, or gray. *G. punctipes* adults are about 3/16 inch (5 mm) long; *G. pallens* is slightly smaller. They may appear as soon as plants emerge, but are most common in midsummer. Both species occur throughout the Western Region, although *G. punctipes* is more common in the deserts while *G. pallens* is more common in the San Joaquin Valley. Nymphs and adults of both species are easily distinguished from lygus and other plant bugs by the broad head with bulging eyes and the short antennae slightly enlarged at

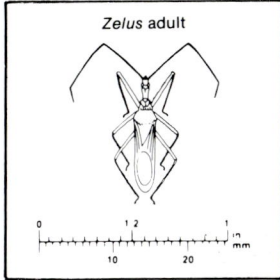

DAMSEL BUG

adult

0   1 2   1 in mm
10   20

ASSASSIN BUG

Zelus adult

0   1 2   1 in mm
10   20

Damsel bugs are slender and usually tan or gray. An adult is at left, a nymph at right.

Assassin bugs such as *Zelus renardii* are among the most obvious predators in cotton fields.

the tips. False chinch bugs, about the same size and color as bigeyed bugs, are more slender.

Eggs of bigeyed bugs are laid on leaves, often in spider mite colonies. They are white to tan with a distinctive red eyespot that develops soon after the egg is laid. Nymphs are light gray to blue. Newly hatched nymphs feed on small, inactive prey such as mite or insect eggs. Older nymphs and adults may also feed on eggs, but they are capable of attacking first-instar caterpillars, lygus nymphs, mites, and other active prey. Bigeyed bugs also obtain food from the host plant, feeding on nectar from extrafloral nectaries and to some extent on plant tissues. Nectar helps sustain populations when prey is scarce.

Development from egg to adult takes 2 to 3 weeks in summer, and a complete generation takes from 17 days to a month or more. Bigeyed bugs pass the winter as adults.

Cotton fields close to alfalfa, an important source of bigeyed bugs, generally have more of them and other predators. Bigeyed bugs also develop on other crops and on weeds. They are attracted to cotton as long as plants are growing; their numbers decline once plants start to cut out or after irrigation is stopped.

Minute Pirate Bugs   *Orius* spp.

Minute pirate bugs are among the first predators to appear in cotton each season. They are attracted to colonies of thrips, their main food. Later in the season, they are concentrated in terminals and flowers, where thrips are most abundant. They also feed on spider mites, whiteflies, aphids, lygus nymphs, insect eggs, newly hatched caterpillars, and other small prey.

Clusters of assassin bug eggs are laid on leaves.

Collops beetles such as *Collops vittatus* feed on insect eggs and other small prey.

The spined assassin bug, *Sinea diadema*, has stiff spines on the front legs to hold its prey.

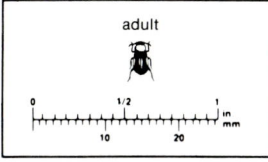

COLLOPS BEETLE

adult

0   1/2   1 in mm
10   20

O. tristicolor is found on cotton in most of the Western Region. *O. insidiosus* occurs in eastern Arizona and New Mexico. Adults of both species are black with three triangular, white or silvery markings on the back. Eggs are inserted in leaves, mostly along veins on the lower surface. Nymphs are usually yellow or orange and have the same predaceous habits as adults. Development from egg to adult takes 10 to 20 days in summer.

Lacewings   *Chrysopa carnea* and others

Although some lacewing adults are predaceous, it is the larvae that are important as predators in cotton. Lacewing larvae are flattened, tapered at the tail end, and have a pair of conspicuous, sickle-shaped mandibles. Active mainly at night, they crawl over plants searching for prey to puncture and drain of fluids with their hollow mandibles. Lacewing larvae attack nearly any prey that is small enough, including spider mites, aphids, whiteflies, insect eggs, and small caterpillars.

The larvae pupate in spherical, white cocoons in sheltered places on the plant. Adults are slender insects ¾ inch (15 to 20 mm) long, with four delicate, net-veined wings and long antennae. They flutter like moths when disturbed. *Chrysopa carnea*, the most common species, is green with golden eyes. It lays eggs singly on threadlike stalks and overwinters as an adult.

Damsel Bugs   *Nabis* spp.

Damsel bugs usually are not abundant in cotton until midseason, when they may become quite numerous. Adults and nymphs feed on aphids, leafhoppers, lygus bugs, and small caterpillars. Eggs are inserted in plant tissue in the same way as are lygus eggs, but the exposed portion is round instead of oval. The two common species are *Nabis americoferus* and *N. alternatus*.

Assassin Bugs   *Zelus* and *Sinea* spp.

Because of their large size, assassin bugs often are the most conspicuous insects in cotton from mid- to late season. Both nymphs and adults will attack nearly any insect prey, but they are relatively ineffective in controlling cotton pests. *Zelus socius* and *Z. renardii* are slender, green species with red wings. *Sinea diadema* and related species are brown and have front legs that are slightly enlarged and spiny. *Sinea* eggs are laid in double rows; each has a cap resembling an inverted mushroom cap.

Collops Beetles   *Collops* spp.

Collops beetles feed on insect eggs, including those of stink bugs and caterpillars, and also on spider mites, aphids, and small caterpillars. *Collops vittatus* is the most common species in most areas. *C. marginellus* is a paler, slightly smaller species found in the desert valleys. Collops larvae are predaceous, but they live mostly on the ground and are seldom found on plants.

Pests that Damage Squares and Bolls

Insects that feed on squares and bolls reduce yield if they are numerous enough and if the injury occurs at a time when plants cannot compensate for it. Plants that lose squares early in the fruiting cycle often retain more of the squares produced later, so the final boll load may be unchanged. If the season is too short, however, a delay in fruiting may not leave time for late bolls to mature. Also, the loss of squares reduces the energy demand that normally restrains vegetative growth. If plants lose too many early squares, energy may be diverted into excessive vegetative growth. Injury to bolls that would otherwise have time to mature subtracts directly from potential yield.

Insect pests of squares and bolls are of two general kinds: piercing and chewing. Insects with piercing-sucking mouthparts (Hemiptera) include lygus bugs, cotton fleahoppers, and stink bugs. They pierce fruiting structures with needlelike stylets, inject a saliva that breaks down plant tissues, and suck out the resulting fluids. Lygus bugs are by far the most widespread and damaging; stink bug injury is usually localized and sporadic, and the cotton fleahopper is significant only in eastern New Mexico.

Insects with chewing mouthparts, including the boll weevil and such caterpillars as the bollworm, tobacco budworm, and pink bollworm, usually create noticeable holes in squares and bolls. The bollworm occurs throughout the Western Region. Pink bollworm and tobacco budworm are generally limited to the desert valleys, although damaging infestations of pink bollworm do occasionally occur in New Mexico. The boll weevil is destructive in parts of Arizona and may have the potential to spread to other areas. Certain caterpillars that are primarily foliage feeders, such as the beet armyworm and the omnivorous leafroller, injure fruit occasionally, but their injury usually is not economically significant.

Bollworm
Heliothis zea
Tobacco Budworm
Heliothis virescens

The bollworm and tobacco budworm are so similar in their biology and injury to cotton that they can be discussed together as *Heliothis*. The bollworm attacks cotton everywhere in the Western Region. The budworm is important as a pest of cotton only in the desert valleys, although it also occurs in small numbers on cotton in eastern New Mexico. The budworm is present in the San Joaquin Valley, but it feeds mostly on ornamental plants and is not known to attack cotton there.

Bollworms and budworms can cause significant losses by feeding on green bolls. Older larvae do most of the damage, but control measures must be aimed at small larvae because large ones are hard to kill. Both species have shown substantial tolerance or resistance to many common insecticides. Treatment thresholds are based on counts of small larvae in the tops of plants.

Losses due to *Heliothis* are greatest where natural enemies have been destroyed by insecticides applied for other pests. Severe outbreaks are most common in the deserts following repeated treatments for pink bollworm. In the San Joaquin Valley and central New Mexico, damaging outbreaks occur only in occasional seasons when conditions are especially favorable or in fields where insecticides are applied at midseason or later for lygus bugs or other pests.

Description

Heliothis eggs, creamy white when laid, develop a dark red to brown ring after 24 hours. They are nearly spherical or hemispherical and have 10 to 15 distinct ridges radiating from the top. Bollworm and budworm eggs are virtually identical, but both differ from cabbage looper eggs, which are sometimes confused with them. Looper eggs have a larger number of much finer ridges and are usually laid on fully developed leaves, while *Heliothis* eggs are usually on young leaves at the top of the plant (Figure 22).

Newly hatched larvae have several rows of dark tubercles along the back, each bearing one or two bristles. These features are readily visible under a hand lens, and they distinguish young *Heliothis* larvae from smaller species of caterpillars such as pink bollworm and omnivorous leafroller. Newly hatched beet armyworms also have tubercles and bristles, but they usually feed in groups on foliage around the egg cluster and are seldom found singly until later, when they can be identified by other characteristics.

Heliothis larvae can best be recognized by numerous tiny spines on large portions of the skin. The spines are much smaller than the bristles on the tubercles, but they can be seen with a 10x hand lens on larvae in the third instar or older. They distinguish *Heliothis* from all other caterpillars commonly found on cotton in the western U.S., including beet armyworms, loopers, and cutworms. *Heliothis* larvae are so variable in color that it is not possible to rely on color to recognize them.

The bollworm cannot be distinguished reliably from the budworm until larvae are in the third instar. The features that separate the two species are shown in the Key to Caterpillars on page 55. The budworm has a toothlike structure on the inner surface of the mandibles that is

M. G. KINSEY

Figure 22. These photos taken with an electron microscope show the features that distinguish eggs of *Heliothis* from cabbage looper eggs. *Heliothis* eggs (left) are more spherical and have thick ridges radiating like ribs from top to bottom. Looper eggs (right) are more flattened and have a larger number of much finer ridges.

lacking in the bollworm, and it has the tiny spines of the skin extending onto the tubercles on top of the eighth abdominal segment; in the bollworm, these tubercles lack spines.

The smooth, brown pupae of *Heliothis* are formed in cells 2 inches (5 cm) or more below the soil surface. The adult bollworm moth has a wingspan of about 1¾ inches (5 cm); the adult budworm is slightly smaller. Adults of the two species are easy to distinguish by color pattern.

Seasonal Development

Bollworms and budworms feed on a variety of crops, ornamentals, and weeds. In areas with mild winters, such as the desert valleys, at least part of the population continues to develop all year, moving from one host to another. Common winter hosts include alfalfa, lettuce, weeds, and ornamentals. In cooler areas, as in the San Joaquin Valley and New Mexico, *Heliothis* pupae pass the winter in the soil.

Cotton is not a strongly preferred host, so the timing of infestations depends partly on the availability of other crops. Corn is the preferred host of the bollworm (also known as the corn earworm and tomato fruitworm). When corn is available, bollworms attack it first, and a later generation moves to cotton after the corn matures. Beans, tomatoes, and safflower are also favored hosts.

Late season pests, both species seldom appear in cotton in significant numbers before peak bloom, and damaging infestations usually begin in August or later. In the desert valleys, there are several generations a year, including three or four on cotton. High populations are favored by periodic summer rains and by the scarcity of natural enemies in fields treated repeatedly for pink bollworm. In the San Joaquin Valley and New Mexico, there are seldom more than two generations on cotton. Each summer generation requires about 710 degree-days (base 60° F).

Heliothis eggs are laid singly near the top of the plant, usually on young leaves or on the bracts of small squares.

Eggs of bollworms and tobacco budworms are usually on young leaves at the top of the cotton plant. Newly laid eggs are white, but they darken to tan before hatching.

Heliothis eggs develop a reddish brown ring a day after they are laid.

Young *Heliothis* larvae feed on small squares. This larva is about the maximum size that can be controlled effectively with insecticides.

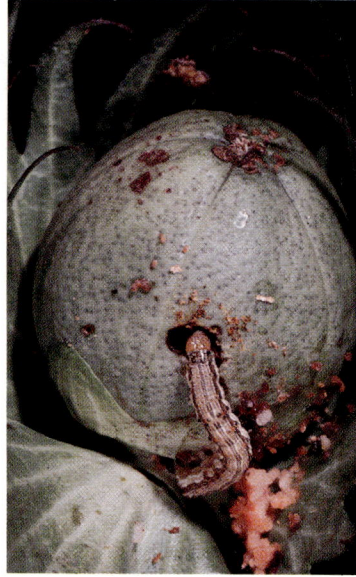

Yield loss due to *Heliothis* results from feeding by older larvae on young bolls. Once larvae reach this size, they are difficult to kill with insecticides.

Upon hatching, the larvae often eat their eggshells, then feed on young leaves for a few hours before moving to small squares.

First and second instars feed mostly on small squares, but they will also feed on vegetative buds. As the larvae grow, they feed on increasingly larger fruiting structures. Fourth and fifth instars feed mainly on large squares, flowers, and green bolls. They often chew a hole in the base of a boll and hollow it out, usually feeding with the posterior part of the body protruding from the hole. There is seldom more than one larva in a single square or boll. Larval development takes about 2 weeks in warm weather and about a month in cool weather.

Mature larvae burrow into the soil to pupate. In spring and early summer, adults emerge from the pupae in about 12 days. In late summer, however, an increasing proportion of the pupae in each generation enter diapause, a state of dormancy in which they pass the winter. Adults from these pupae emerge in spring. In mild winters in the desert valleys, less than half of the population enters diapause; most of it continues development through the winter, although at a slower rate than in summer. In cooler areas, most or all of the population enters diapause.

Adult moths emerge from the pupae at night and climb to the tops of plants where they dry their wings. All adult activity, including flight, mating, egg laying, and feeding, occurs mostly at night. Males respond to a pheromone released by females, and mating usually occurs within 48 hours after emergence. Egg laying usually begins within 72 hours.

Heliothis adults live longer and females lay more eggs when nectar and moisture are available. They obtain nectar from the extrafloral nectaries of cotton plants as well as from other sources. Adults live for 10 to 14 days in summer and up to a month in cool weather. Each female may lay more than 1000 eggs. Cotton plants are most attractive to egg-laying females during flowering and when vegetative growth is lush, especially just after an irrigation.

Heliothis infestations are usually spotty, even in seasons when the general population level is high. They usually start in "hot spots" that increase in size with each generation. Adjacent fields that are otherwise similar often have very different levels of infestation.

Damage

Most losses due to *Heliothis* result from damage to bolls. Damaged bolls usually have a round hole near the base and one or more hollowed out locks. The hole may be up to ¼ inch (6 mm) in diameter, and moist frass usually accumulates around the base of the boll. In additon to feeding internally, larvae may chew shallow gouges in the boll surface. With high humidity in the canopy, injured bolls often rot. In drier situations, they may dry out and remain on plants as brown "mummies."

Squares injured by *Heliothis* usually have a round hole near the base. A small amount of webbing is often present, especially on small squares injured by young larvae. Injury to pinhead squares may not be obvious, as the squares are hidden by foliage and larvae may completely remove them. However, larval frass and the flaring of bracts on larger squares are usually apparent by the time larvae reach the second instar.

You can easily distinguish *Heliothis* injury from other insect damage. Lygus bug injury to squares differs in that there is no entry hole and no evidence of chewing. Pink bollworms never hollow out large portions of bolls and they do not leave frass outside injured bolls; their injury is concentrated in the seeds. Beet armyworm damage is usually associated with feeding on bracts and foliage. Squares infested with boll weevil larvae usually drop with the larvae still inside, and weevil-infested bolls have a distinctive feeding chamber among the seeds. Punctures and frass left by adult boll weevils are also distinctive.

Management

The impact of a *Heliothis* infestation depends on the number of larvae present, the age of the larvae, and the timing of damage relative to the crop's fruiting cycle. Fifth-instar larvae are the most destructive; they not only damage more fruit than do earlier instars, but they damage larger fruit that are harder for the plant to replace. Although large larvae do most of the damage, it is not possible to kill a significant proportion of them once they are older than the third instar. Monitoring and control must therefore be aimed at the eggs and small larvae.

Natural enemies control *Heliothis* in many cases, especially in the San Joaquin Valley. Damaging populations usually do not appear until late in the season, after treatments for other pests have disrupted natural controls. Insecticides are needed only if the population exceeds the treatment threshold while the crop has a significant number of squares or green bolls that will have time to develop into mature bolls by season's end. There is no need to treat once bolls begin cracking, because most bolls are too mature by that time to be susceptible and squares still present will not have time to mature. The same principle applies to long-season desert valley crops, except that there are two periods when injury can occur—one in each fruiting cycle.

Biological Control. The most important natural enemies of *Heliothis* are such predators as bigeyed bugs, minute pirate bugs, damsel bugs, and lacewing larvae. They feed on eggs and small larvae, killing them before they reach the more damaging later instars. Other predators, including assassin bugs, spiders, and collops beetles, also attack *Heliothis* but they have less effect on populations. Studies in the San Joaquin Valley have shown that, due

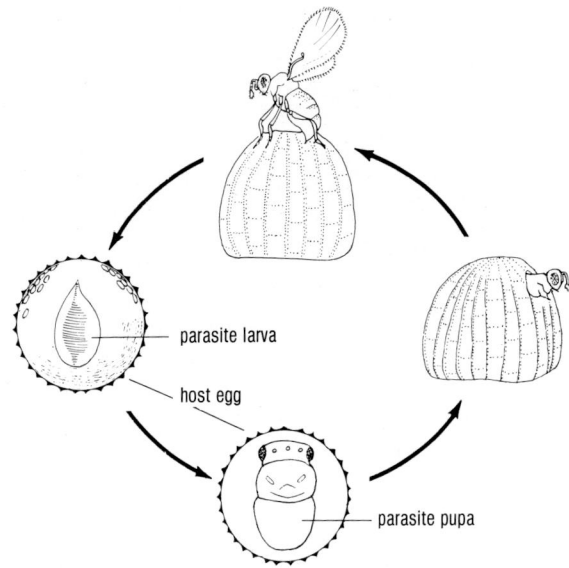

Figure 23. The tiny *Trichogramma* wasp attacks eggs of caterpillars. Larval and pupal stages take place entirely within the egg of the host, and the life cycle may take only a week. Each female wasp attacks many host eggs.

largely to predators, only a small percentage of bollworm larvae survives to the fourth instar in untreated fields.

Parasites of *Heliothis* include *Trichogramma* sp., a tiny wasp that attacks the eggs (Figure 23). Natural populations of *Trichogramma* destroy most *Heliothis* eggs in some fields, especially in untreated desert valley fields. *Trichogramma* is also reared commercially and sold for control of pest caterpillars. Experimental releases have reduced pest egg numbers significantly in some cases, but techniques for using *Trichogramma* on a commercial scale in cotton are still under study. Consult an Extension agent or a commercial supplier for more information. *Chelonus insularis* (*C. texanus*), a parasitic wasp common through the cotton belt, lays its egg in the egg of a bollworm, budworm, or other host, and the parasite larva later kills the host larva. *Hyposoter exiguae*, another wasp, usually kills larvae in the third instar.

BOLLWORM AND TOBACCO BUDWORM

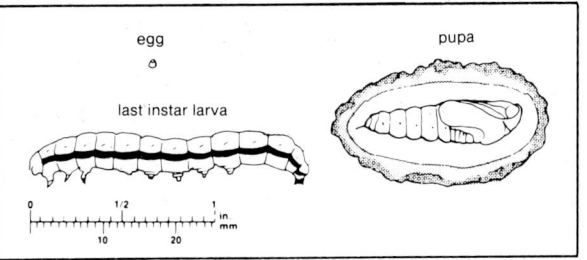

SEQUENTIAL SAMPLING: BOLLWORM LARVAE
SAN JOAQUIN VALLEY

threshold = 20 larvae/100 plants   error rate = 20%
(untreated fields) or 15/100 (treated)   Each Sample is 5 adjacent plants.

sample no.	don't treat*		no. of larvae	treat*	
		*		*	
1	–	–	___	–	–
2	–	–	___	–	–
3	–	–	___	–	–
4	–	–	___	–	–
5	–	–	___	–	–
6	2	3	___	10	8
7	3	4	___	11	9
8	3	5	___	13	10
9	4	6	___	14	11
10	4	7	___	15	12
11	5	7	___	16	13
12	6	8	___	17	14
13	6	9	___	18	15
14	7	10	___	20	15
15	8	11	___	21	16
16	8	12	___	22	17
17	9	12	___	23	18
18	9	13	___	24	19
19	10	14	___	25	20
20	11	15	___	26	21

*Use inner columns (*) for fields that have not been treated. Use outer columns for fields already treated.

Figure 24. Sequential sampling card for bollworm in the San Joaquin Valley. For each sample, count small bollworm larvae on the upper portions of five adjacent plants. Check six samples, or 30 plants, to reach a treatment decision. If there is no decision after 20 samples, sample again in 2 or 3 days.

Many natural enemies of budworms and bollworms breed in alfalfa, sorghum, and other crops as well as in cotton, and they often are more abundant in fields bordered by such crops. In alfalfa fields adjacent to cotton, border harvesting (page 64) helps to maintain a reservoir of natural enemies that can move into the cotton when host populations build up. Where broad-spectrum insecticides are applied repeatedly, predators and parasites usually are not numerous enough to keep *Heliothis* populations below damaging levels.

Cultural Practices. The length of the growing season affects losses due to *Heliothis*. Because populations seldom reach damaging levels before late summer, the greatest potential for injury is in the desert valleys, where budworm damage is often extensive in the second fruiting cycle. In some parts of the cotton belt, such as Texas, losses due to budworm have been greatly reduced by switching to early maturing varieties and a shorter growing season.

Irrigation and fertilization affect *Heliothis* populations by changing the attractiveness of plants to egg-laying females. Rank, lush plants are more attractive, and fields with large areas of rank growth may have more damage than surrounding fields with normal growth. Keep water, fertilizer, and plant density at recommended levels to prevent this.

Monitoring. Each area in the Western Region has its own treatment threshold based on the number of *small* larvae—less than about ⅜ inch (1 cm) long—per 100 plants. Treatments are aimed at small larvae because larger ones are hard to control and because it is essential to destroy the larvae before they reach the more damaging later instars.

Monitoring involves searching for small larvae on foliage and squares at the tops of randomly chosen plants. Although larvae may be present anywhere on the plant, small larvae are concentrated near the top, where most eggs are laid. On each plant sampled in monitoring, look carefully for larvae on leaves, squares, and other growth from the top of the plant down to the fourth node below the newest partly unfurled leaf (Figure 5, page 12).

Methods recommended for choosing sample plants differ among areas. These differences are not crucial; adapt suggested sampling plans to fit your needs. It is important, however, to follow the general directions on page 35 to make sure sample plants are chosen randomly. Because *Heliothis* infestations often begin at the edge of a field and in areas with rank growth, check these areas regularly. Decisions for treating a whole field, however, should be based on random samples of plants taken throughout the field. Don't limit sampling to known "hot spots" unless you plan to treat these areas separately.

While searching for larvae, watch for bigeyed bugs, minute pirate bugs, damsel bugs, and lacewing larvae. You need not count them, but the presence or absence of these predators can help you make treatment decisions when the *Heliothis* population is close to the threshold.

Pheromone traps can alert you to periods when adult moths are active. Set traps out in mid-July and follow manufacturer's directions for placement and handling. When adults appear in traps consistently for several nights, begin monitoring the field regularly. Never schedule insecticide applications on the basis of trap catches; without field sampling, there is no way to know whether moths are laying eggs on the crop.

San Joaquin Valley. There are two treatment thresholds for bollworms in the San Joaquin Valley. In fields that have *not* been treated with broad-spectrum insecticides, treat when there are 20 small bollworms per 100 plants. In fields that *have* been treated previously, treat when there are 15 small bollworms per 100 plants.

Begin sampling plant terminals for bollworms in the first week of August and continue until most bolls have matured. The sampling method recommended in the San Joaquin Valley is to check five adjacent plants at each stop as you pass through the field. Choose the first plant at random; then check its mainstem terminal and those of the four plants next to it. Checking five plants each time reduces the walking needed between samples plants, and it helps to reduce the chance that samples will contain too many tall plants.

For sequential sampling, use the sample card in Figure 24. The two sets of boundary columns on the card represent the two treatment thresholds. The outer columns are for previously treated fields, and the inner columns are for untreated fields. A minimum of six samples is needed to reach a treatment decision; this means checking the terminals of 30 plants. With a standard sample plan, check at least five samples of five plants in each quarter of the field—a total of 100 plants per field. If the field is larger than about 80 acres, divide it into more than four areas and take a set of five samples in each area. Calculate the number of small larvae per 100 plants to decide on treatment.

Because damaging bollworm infestations do not occur regularly in the San Joaquin Valley, a complete sampling program is not always needed in every field. One approach is to check field edges and areas with rank growth each week until you find bollworm eggs or larvae; then begin routine sampling. Always take a complete set of samples before deciding whether treatment is needed.

Desert Valleys. Bollworm and tobacco budworm often occur together in desert cotton fields. As it is not possible to separate the two species while they are small enough for insecticides to be effective, you must treat them alike in making decisions. The treatment threshold is 10 to 12 small larvae per 100 plants.

Begin sampling in mid-July or about 1 to 2 weeks after peak squaring; continue sampling weekly until most bolls have matured. In crops with a second fruiting cycle, continue until "top crop" bolls have matured. Don't treat for *Heliothis* during the cutout period when there are few susceptible squares or bolls.

For standard sampling, check for larvae on the terminal growth of at least 100 plants chosen at random. Divide fields of up to 80 acres into quarters and check 25 plants in each quarter. Divide larger fields into more areas and check 25 plants in each area. Use Figure 25 for sequential sampling.

Pheromone traps are available for bollworm and tobacco budworm. If you use both kinds of traps, keep them separated by at least 200 yards (180 m) so that they will not interfere with each other. Set out traps by first bloom and follow the manufacturer's recommendations. Pheromone lures for *Heliothis* are still being perfected and interpretation of trap results is still under study.

SEQUENTIAL SAMPLING: BUDWORM/BOLLWORM LARVAE DESERT VALLEYS

threshold = 12 larvae/100 plants
error rate = 20%

plant no.	don't treat	no. of larvae	treat
1	—	___	—
2	—	___	—
3	—	___	—
4	—	___	—
5	—	___	—
6	—	___	—
7	—	___	—
8	—	___	—
9	—	___	—
10	0	___	4
11	0	___	4
12	0	___	4
13	0	___	4
14	0	___	5
15	0	___	5
16	0	___	5
17	0	___	5
18	1	___	5
19	1	___	5
20	1	___	5
21	1	___	6
22	1	___	6
23	1	___	6
24	1	___	6
25	1	___	6
26	1	___	6
27	1	___	6
28	1	___	7
29	2	___	7
30	2	___	7
31	2	___	7
32	2	___	7
33	2	___	7
34	2	___	8
35	2	___	8
36	2	___	8
37	2	___	8
38	2	___	8
39	2	___	8
40	2	___	8

Figure 25. Sequential sampling card for bollworm and tobacco budworm in the desert valleys. For each sample, count small bollworms and budworms on the upper portion of one plant chosen at random. Check at least 10 plants to make a treatment decision. If there is no decision after checking 40 plants, sample again in 2 or 3 days.

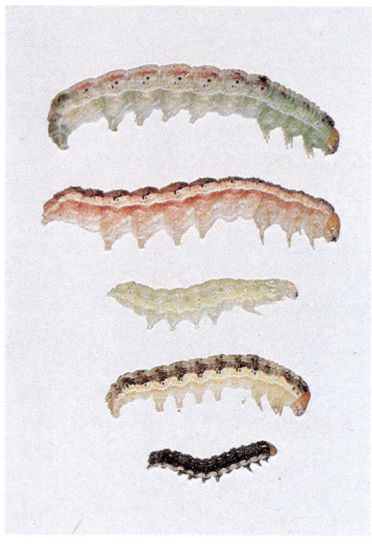

Heliothis larvae have numerous tiny spines on most parts of the skin, in addition to the dozen or so longer bristles found on each segment. The spines are visible under a hand lens.

Bollworms and tobacco budworms vary so greatly in color that you cannot rely on color to distinguish them from other caterpillars.

Bollworm moths mating at night near the top of a cotton plant.

A mating pair of tobacco budworm adults.

Bigeyed bugs are important natural enemies of bollworms and tobacco budworms. This nymph is feeding on a bollworm egg.

The tiny parasitic wasp, *Trichogramma pretiosum*, lays its egg in the egg of a bollworm.

New Mexico. The treatment threshold in New Mexico is 6 to 10 small bollworms per 100 plants. Use the higher threshold for untreated fields early in the season and the lower one for fields already treated with insecticide. Begin monitoring on July 1 and continue until about 10% of the bolls have cracked. Sampling procedures are the same as in the San Joaquin Valley; for sequential sampling, use Figure 26. To use pheromone traps, set them out at first bloom and follow the manufacturer's directions.

Control with Insecticides. Control of budworms and bollworms is complicated by their tolerance or resistance to many common insecticides, including organophosphates and carbamates (Table 8). The budworm is generally more tolerant than the bollworm. Larvae of either species are difficult to control with any material now available once they reach the third instar—that is, when they are more than about ⅜ inch (1 cm) long—so applications must be aimed at smaller larvae.

There is evidence that the tobacco budworm is developing significant resistance to the widely used synthetic pyrethroids (Figure 27). The resistance is based on decreased nerve sensitivity. This is the same kind of resistance that bollworms, budworms, and many other insects have to chlorinated hydrocarbons such as DDT. This kind of resistance is associated with a high degree of cross-resistance. Once resistance to a single compound develops, it quickly broadens to make the insect resistant to similar compounds. While most current evidence of resistance to pyrethroids involves permethrin (Pounce/Ambush), it is probable that the budworm will eventually become resistant to all synthetic pyrethroids. A closely related species, *Heliothis armigera*, is an important cotton pest in Australia and has become highly resistant to all pyrethroids.

Development of a high level of resistance to pyrethroids will greatly limit the choice of chemical controls. The only other insecticide known to be effective is chlordimeform (Galecron/Fundal), which kills first instar larvae as they hatch. Microbial insecticides, such as *Bacillus thuringiensis* (Dipel/Thuricide) and *Baculovirus heliothis* (Elcar), have been used successfully against *Heliothis* in some cases, but the degree of control has been too variable for these materials to be relied upon as the sole means of control in commercial fields. Research is continuing, however, so these insecticides may play a greater role in the future.

To slow development of pyrethroid resistance, avoid exposing pest populations to pyrethroids unnecessarily. Save pyrethroids for cases where sampling shows they are

Figure 27. Resistance to permethrin in tobacco budworm populations in the Imperial Valley, California, as measured by the LD_{50}—the amount of pesticide needed to kill 50% of a sample of larvae. The dose required to kill 50% in 1981 was more than 60 times greater than in 1975. Data courtesy of J. L. Martinez-Carrillo and H. T. Reynolds.

SEQUENTIAL SAMPLING: BOLLWORM LARVAE NEW MEXICO

threshold = 10 larvae/100 plants (early season, untreated fields) or 6 larvae/100 (late season, treated fields)   error rate = 20%

sample no. (5 plants)	don't treat*		no. of larvae	treat*	
		*			*
1	–		_____	–	–
2	–	–	_____	–	–
3	–	–	_____	–	–
4	–	–	_____	–	–
5	–	–	_____	–	–
6	0	1	_____	6	5
7	1	1	_____	7	5
8	1	2	_____	8	5
9	1	2	_____	8	6
10	1	3	_____	9	6
11	1	3	_____	9	7
12	2	3	_____	10	7
13	2	4	_____	11	7
14	2	4	_____	11	8
15	2	5	_____	11	8
16	2	5	_____	12	8
17	3	5	_____	13	9
18	3	6	_____	14	9
19	3	6	_____	14	10
20	3	7	_____	15	10

*Use inner columns (*) for untreated, early season fields. Use outer columns for late season fields and fields already treated.

Figure 26. Sequential sampling card for bollworm in New Mexico. Each sample is five adjacent plants. Sampling procedure is the same as for the San Joaquin Valley.

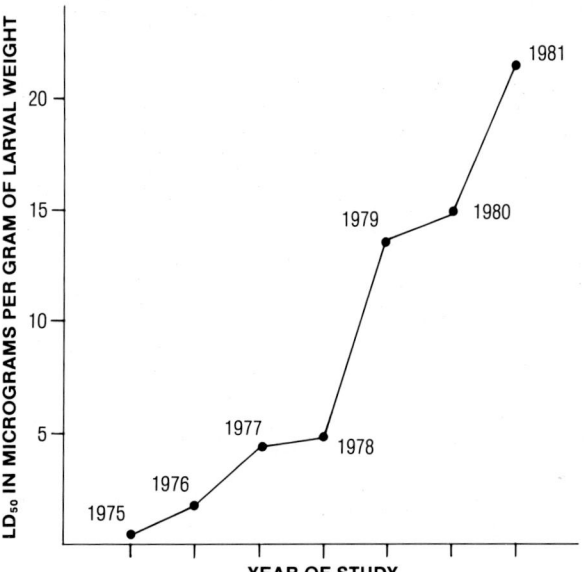

definitely needed to control a damaging *Heliothis* population. Don't use them for pests such as pink bollworm that can be controlled with other insecticides. When you do need a pyrethroid, remember that it may promote an outbreak of spider mites.

Due partly to destruction of natural enemies, *Heliothis* populations often rebuild quickly following applications of broad-spectrum insecticides. Once you have applied an insecticide, additional treatments may be needed later. If you begin treatment too soon, before the crop has reached the susceptible stage or before the population reaches the treatment threshold, resurgence of the *Heliothis* population may result in greater losses and higher control costs later.

Although insecticides are aimed mainly at young larvae, most insecticides also affect adults. If possible, treat at night, when adults are active. Night application also reduces injury to bees.

Pink Bollworm
Pectinophora gossypiella

Pink bollworm has been the key pest of cotton in Arizona and southern California since the mid-1960s. It infests most fields there every year, and insecticides used for control often promote costly outbreaks of tobacco budworm, bollworm, cotton leafperforator, and other secondary pests. In New Mexico, pink bollworm is significant only in occasional seasons.

Pink bollworm is not widely established in the San Joaquin Valley, although adults are apparently carried there by winds from southern California. Small numbers of larvae are found in the Valley occasionally. Timely plowdown in compliance with local regulations prevents wider infestation. Release of millions of sterilized adults each season by the USDA in cooperation with other public agencies is designed to prevent immigrating adults from mating and reproducing.

Shredding and discing crop debris promptly after harvest reduces overwintering populations to a level that can be managed economically. Insecticides are needed only when *both* a damaging population *and* a significant number of susceptible bolls are present. Time applications by picking samples of bolls and checking them for larvae. Use pheromone traps and degree-day models for monitoring population cycles, but do not rely on them for scheduling treatments.

Description

Adults, mottled gray and brown moths, are active at night and hide in sheltered places during the day. Eggs are seldom seen in the field because they are small and are usually hidden under the calyx of young bolls. Eggs are about 1/50 inch long and 1/100 inch in diameter (0.5 by 0.25 mm) and are laid singly or in small groups. Newly hatched larvae are white and about 1/25 inch (1 mm) long. In the third instar, they begin to show a pink color pattern that deepens to reddish pink in the final instar. The pupa is formed in a silk cocoon on the ground, in the soil, or inside damaged bolls.

Seasonal Development

Of the pink bollworm's three to five generations a year, the first one or two feed mainly in squares and flowers in spring; later generations feed in bolls. Populations reach a peak in August or September, usually in the fourth or fifth generation. Larvae that overwinter in diapause produce adults that are the source of new infestations in spring (Figure 28). Pink bollworms feed to a limited extent on certain weeds and ornamentals, mostly those in the same family as cotton; okra is the only crop host other than cotton in the U.S.

In the low deserts, adults emerge from overwintering larvae as early as February. Some overwintering moths may emerge as late as August, but peak emergence usually occurs from April to early June. At higher elevations, such as Safford, Arizona, emergence may not begin until April. Accumulation of degree-days starting on January 1 can predict the spring emergence peak (Table 9).

Table 9. Degree-Days Required for Spring Emergence and Summer Generations in Pink Bollworm, Calculated According to Three Models Used in the Western Region.

Model: developmental threshold/upper temperature limit	Degree-days from January 1 required for spring emergence: begin	peak	end	Degree-days required for one summer generation
60° F/no upper limit	200	675	1100	967
55° F/86° F	500	1180	2200	800
55° F/no upper limit	500	875	2250	750

Some overwintering adults usually emerge before cotton begins squaring, laying eggs on leaf buds and stems, where larvae rarely survive. Females that emerge later lay eggs on squares, where the larvae feed mainly on the anthers. Squares must be at least 10 days old to support a larva. Larvae in squares often fasten the petals together with silk, creating a "rosetted" bloom that fails to open normally. Some infested squares may drop, but most squares attacked by pink bollworm develop into normal bolls.

The population increases slowly while larvae are feeding in squares, but it increases rapidly as soon as green bolls are available. Bolls of upland cottons are most susceptible to pink bollworm from 14 to 21 days after flowering. Older bolls are so tough that few larvae can penetrate them. Pima bolls are relatively soft and are susceptible up to about 30 days after flowering.

Eggs laid on bolls are usually placed under the calyx. After hatching, the larvae bore into the boll almost immediately without eating any of the surface tissue, so their exposure to insecticides and predators is very brief. The small entry hole is rarely visible from outside, but a round, swollen "wart" usually develops around it on the inner surface of the boll wall. Warts may not form, however, in very young bolls. In older bolls, larvae often tunnel beneath the boll wall's inner surface before entering the lint, leaving a distinctive mine.

Larvae cut and stain the lint as they work their way to the seeds, where they do most of their feeding. On average, each larva damages four to five seeds. After passing through four instars, the larvae usually make a round exit hole and drop to the ground to pupate.

In midsummer, most larvae pupate immediately after feeding. Adults from these "short-cycle" larvae emerge in about 11 to 17 days. In late summer, however, more and more larvae in each generation enter diapause and do not pupate until spring. These "long-cycle" larvae may either drop to the ground or remain in seeds; they pass the winter in a silk cocoon. The proportion of long-cycle larvae increases sharply in mid-September, and nearly all larvae enter diapause by early October. Usually, however, a small number of larvae continues feeding as long as food is available and there is no frost, so bolls and squares left on plants over the winter can be heavily infested. After plowdown, the entire population consists of diapausing larvae.

Adults mate at night at the tops of plants, usually within 1 or 2 days after emerging. Females release a pheromone that attracts males for mating. In summer generations, females lay about 200 eggs over several nights. They live longer and lay more eggs when moisture or sugar is available from extrafloral nectaries.

As long as shelter and egg-laying sites are available, adults do not move far from where they emerge. Early in the season, however, when plants are too small to provide shade, the moths may move to nearby crops or other vegetation for shade and moisture. They also tend to fly later in the season when populations are high and few young bolls are available. Although they are weak fliers, adults may be carried long distances by wind. They commonly move 100 miles (160 km) or more from the Coachella Valley to the San Joaquin Valley.

Damage

The effect of pink bollworm injury on lint yield and quality depends on humidity as well as on the number of larvae present. Under dry conditions, no measurable

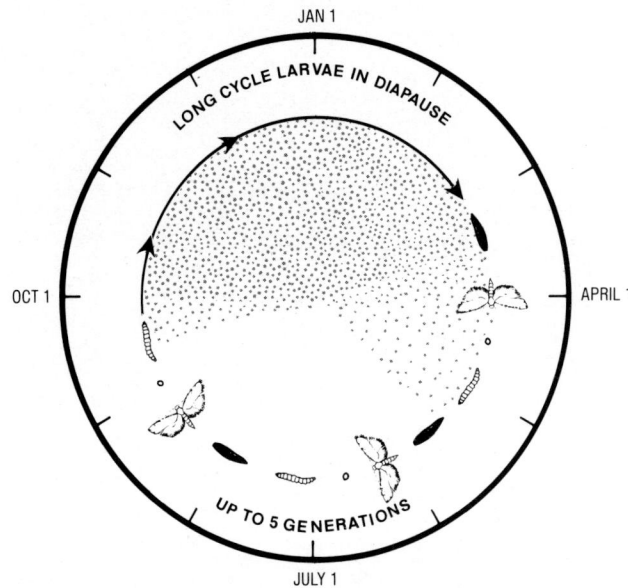

Figure 28. Pink bollworms pass the winter as long-cycle larvae in diapause. These larvae pupate in spring, and the adults that emerge begin a series of summer generations. The exact timing of events varies according to degree-day accumulations in each area.

reduction in yield occurs until 25 to 30% of the bolls are infested; at this level a significant proportion of the infested bolls have more than one larva. With high humidity, however, it takes only one or two larvae to destroy an entire boll, because damaged bolls are vulnerable to infection by fungi that cause boll rots. These fungi include the one that produces aflatoxin, so damaged bolls are more likely to have a high level of aflatoxin contamination.

You can recognize pink bollworm injury by the damaged seeds, stained lint, warts or mines on the inner boll wall, and exit holes left by mature larvae. Stink bugs also stain lint and produce warts, but the stain is usually more extensive and the warts, irregular in shape, are often clustered. Pink pollworms never hollow out large portions of the lint or leave frass outside the boll as do *Heliothis* larvae. Pink bollworm diapause and pupation shelters, sometimes found inside infested bolls, have a white silk lining; chambers constructed by boll weevil larvae do not.

Management

The objectives of pink bollworm management are to keep infestations below damaging levels in the current season—without creating secondary outbreaks of other pests—and to reduce the overwintering population that will threaten the following season's crop. The main control tools are plowdown and insecticides. Mass trapping, mating disruption, and resistant varieties may sometimes be useful.

A round wart usually develops on the inner side of the boll wall where a pink bollworm larva has entered.

E. SEALE

Pink bollworms sometimes tunnel in the boll wall before entering the lint, creating a distinctive mine on the inner surface.

Except for the brown head capsule, small pink bollworms are the same color as lint. Look closely for them when cracking bolls in monitoring.

Older pink bollworms stain lint around the seeds.

Because of the danger of secondary outbreaks, especially in the low desert valleys, it is essential to limit insecticide treatments to those periods when susceptible bolls are present and when sampling shows the percentage of infested bolls is above the treatment threshold. It is rarely necessary to use insecticides to kill moths from the overwintering population or the first field generation that develops in squares and flowers. In most cases, treatment need not begin before emergence of the second field generation—the first generation to develop in bolls.

Cultural Controls. Cultural controls aim to keep the number of larvae that survive between seasons as low as possible. The population then takes longer to build up to damaging levels the following season, so the need for insecticides is delayed. Cultural controls applied over a large area have virtually eliminated pink bollworm as a major pest in parts of Texas where damage once was severe. Principal cultural controls are early termination of the crop and prompt plowdown after harvest.

Buildup of the overwintering population occurs in September, when long-cycle larvae are entering diapause. The best way to reduce winter survival is to eliminate the food supply for these larvae by cutting off irrigation early enough to stop production of green bolls by early September. In most soils, this means stopping irrigation by mid-August; continuing irrigation into September usually provides enough late green bolls to produce a large overwintering population. Early harvest results in less than the maximum possible yield, but it can be cost effective because of savings on water, insecticide, and other expenses.

Researchers have tested growth regulators that selectively stop growth of young fruiting structures while permitting older fruit to mature. These materials may have the potential to limit the food supply for overwintering pink bollworms with less effect on yield than an early irrigation cutoff, but growth regulators are not now registered for this purpose.

Regardless of when you terminate the crop, plow down promptly after harvest. Any postharvest cultivation helps reduce populations, but shredding cotton stalks and discing are essential. Shredding destroys many larvae directly, and it permits uniform burial of debris so that moths cannot escape in spring. Nearly all larvae are in the upper 2 inches (5 cm) of soil. Cross-discing or plowing to a depth of at least 6 inches (15 cm) ensures that few moths will emerge.

The worst practices from the point of view of pink bollworm control are the stubbing or abandonment of cotton fields. In stubbed fields, there is no way to destroy unharvested bolls and other debris that harbors overwintering larvae. Cotton left in the field over the winter provides late bolls for the development of long-cycle larvae and produces fruit early in spring, allowing the population to begin building earlier than would be possible in planted cotton. Stub cotton serves as a source of infestation for planted cotton; fields near stub cotton may be infested 3 or 4 weeks earlier than are other planted fields.

A delay in planting may increase the proportion of the overwintering generation that emerges before squares are available. In most areas where pink bollworm occurs, the optimum period for planting is about 1 month long. Planting toward the end of this period can delay the appearance of squares and green bolls as much as 2 weeks without affecting yield. Such a delay can significantly reduce the ability of the overwintering population to reproduce. Use degree-days to judge whether a delay in planting would allow the crop to escape a significant part of the overwintering emergence.

Rotation crops may also affect pink bollworm. Winter crops may reduce the number of adults emerging in spring, but shading in small grains and the higher moisture level in irrigated crops will lower soil temperature. Cooler soil may delay emergence, sometimes increasing the proportion of adults that emerge after squaring begins. Summer fallowing, rotating with other summer crops, or planting alfalfa for 2 or more years will greatly reduce populations present when cotton is replanted.

Biological Control. Pink bollworms are exposed to predators and parasites only for a short part of their life cycle. Eggs and first instars are hidden under the calyx of bolls, so natural enemies must search to find them. Larvae in bolls are exposed only when they leave the bolls to pupate. Larvae in squares and flowers are more vulnerable.

Predators such as lacewing larvae and bigeyed bugs feed on pink bollworm eggs and small larvae, but the large numbers that would be required for reliable control are seldom found in cotton fields, especially in fields treated with insecticide.

The most effective native parasite is *Bracon platynotae*, a small wasp. It attacks pink bollworms in squares, but has little impact on those in bolls. Parasites imported from other parts of the world have in some cases reduced pink bollworm numbers where they were released, but establishing them permanently is difficult. Controls directed at pink bollworm also destroy the parasites, and the parasites die out in winter, when host larvae are not available.

Monitoring. Sampling bolls is the only reliable way to judge the extent of an infestation. Treatment thresholds are expressed in terms of the percentage of infested bolls found in samples.

While not adequate for scheduling insecticide treatments, pheromone traps and degree-day accumulations can help concentrate boll sampling when populations are increasing. They can also help in making treatment decisions when sampling shows the population is near the treatment threshold. For example, if trap catches

Pink bollworms feeding in flowers early in the season fasten petals together with silk, creating a "rosetted" bloom.

Mature pink bollworm larvae spin a silk cocoon on the ground or inside infested bolls. Short-cycle larvae soon pupate and emerge as adults, but long-cycle larvae remain in the cocoon over the winter before pupating.

A pink bollworm moth.

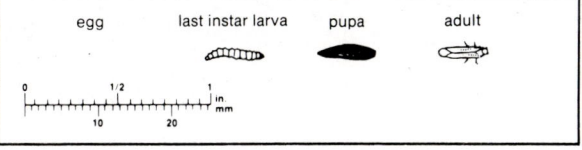

PINK BOLLWORM

| egg | last instar larva | pupa | adult |

0   1/2   1 in
10   20   mm

SEQUENTIAL SAMPLING: PINK BOLLWORM

threshold = 10% infested bolls
error rate = 20%

bolls sampled	tally per 10 bolls	don't treat	total tally	treat
10	___	—	___	—
20	___	—	___	—
30	___	—	___	—
40	___	2	___	6
50	___	3	___	8
60	___	4	___	9
70	___	5	___	10
80	___	6	___	11
90	___	6	___	12
100	___	7	___	13

Figure 29. Sequential sampling card for monitoring pink bollworm infestation in bolls. For each 10 bolls, keep a tally of the number infested in the first column of blanks. Transfer results to the center column to keep a running total. You must check at least 40 bolls to reach a treatment decision. If there is no decision after 100 bolls, sample again in 2 or 3 days.

and degree-days indicate that the population peak has passed, you may then be able to delay a treatment until the population builds up again in the next generation. On the other hand, if the population peak is yet to come, a treatment may be needed right away. A convenient "slide rule" has been developed for matching pink bollworm population cycles to cotton development in the desert valleys. The slide rule publication is listed in the References.

Boll Sampling. In picking boll samples, choose bolls 14 to 21 days old, the stage when they are most likely to be infested. Bolls at this stage are fully grown, but have only partially developed lint. They feel slightly spongy but firm when squeezed between the thumb and forefinger. Younger bolls are watery, while older bolls are dry and do not yield when squeezed.

Crack the bolls open and check each lock for larvae and for warts or mines where larvae have entered. Young larvae are usually in the lint close to a wart or mine, nearly always in the same lock where they entered. Except for the brown head, small larvae nearly match the color of lint, so look carefully for them. Older larvae, usually near the seeds, are easier to find because they stain the lint around the feeding site. Don't count a boll as infested unless you find a larva, even if there is a wart or mine in one lock. It is important to find the larvae while they are small so there will be time to kill the adults early enough in the generation to prevent the level of boll infestation from going too high.

For standard sampling, pick 25 bolls at random in each quarter of the field. If the field is larger than about 80 acres, pick one sample of 25 bolls in each 20 acres. A sequential sampling plan (Figure 29) has been designed for sampling bolls in Pima cotton in Arizona. The plan could also be used in upland cotton, although it has not been tested extensively in upland cotton.

Pick bolls as you walk through the field, taking only one or two from each plant. Choose bolls randomly; don't pick them at the edge of the field and don't limit sampling to areas where growth is rank. You do not need to crack the bolls in the field; it may be more convenient to put them in paper bags and check them later. Label each sample with a field number or other identification.

Degree-Days. To follow population cycles, keep a record of degree-days starting on January 1 and continuing through the season. Use the degree-day figures to predict the emergence of adults from overwintering larvae. You may be able to improve the accuracy of predicting field generations by starting a new degree-day accumulation when the first squares are 10 days old; this is when development of the first field generation begins. Use the number of degree-days required for a generation (Table 9) to estimate how long it will take for the first field generation to emerge after the first 10-day-old squares have appeared.

Pheromone Traps. Pheromone traps provide an index of adult activity. If adults appear in traps while susceptible bolls are present, you should be sampling bolls to monitor the level of infestation. Trap results also help to confirm whether a population is increasing or decreasing, although they cannot precisely identify population peaks. Pheromone traps occasionally attract moths other than pink bollworms, but only rarely in numbers large enough to affect management decisions.

Most widely used for pink bollworm is the delta trap, a folded cardboard shelter with a sticky inner surface to hold the moths. The much larger cone traps are now used relatively little. Oil traps, used mostly in Pima cotton in Arizona, consist of a paper cup with a smaller plastic cup inside. The paper cup has holes in the sides to admit the moths, and the plastic cup holds a light grade of mineral oil; vegetable oil cannot be used because it attracts animals that disturb the traps. Each oil trap can hold up to 2,000 moths, many more than other traps. Because of this large capacity, oil traps can be used for early season mass trapping as well as for monitoring. Several pheromone formulations are available commercially for use in traps; follow the manufacturers' directions for using them.

Set out pheromone traps at first square and maintain them throughout the season. Use one for each 20 acres, with a minimum of two per field. Set them at least 50 paces from the end of a row and 50 rows from the edge of the field. Traps at the edges do not yield reliable results (Figure 30).

Attach traps to stakes in such a way that you can move them up to keep them even with the top of the canopy as the crop grows. Traps more than a foot above the canopy or hidden below it catch fewer moths. If you set out traps while plants are still seedlings, place them within a foot of the bed surface. Delta traps work best when kept horizontal and when the long axis of the trap is parallel to the wind direction. Mark them clearly and instruct tractor drivers to move the stakes to avoid knocking them over during cultivation.

Check delta traps every 1 to 3 days. Replace them whenever they have caught a total of 100 moths and whenever the average catch per trap is 25 moths or more in one check period. Also, replace them any time the sticky surface gets dirty, even when no moths have been caught. Check oil traps every 3 or 4 days; pick the moths out with forceps or strain the oil to remove them.

Record the number of moths in each trap and calculate the average number per trap for each field. Keep a record for each field showing the average number of moths per trap per night (Figure 31).

You can compare current trap results with those from previous seasons if you have charted catches according to degree-days. For example, you could compare the average trap catch at 675 °D with the average catch at 675 °D in previous years. If a series of such comparisons shows that catches are consistently higher or lower than before, expect more or less population pressure. Comparisons of trap catches by calendar date are not reliable because weather influences emergence. Be sure to compare catch results from the same kind of trap, the same pheromone lure, and the same location each year.

Control. Pink bollworm eggs and larvae are seldom exposed where insecticides can reach them, so applications must be aimed at adults. Applications are most effective at night, when adults are flying and mating at the tops of the plants.

Several insecticides, including certain organophosphates, carbamates, and synthetic pyrethroids, are effective. Populations have shown tolerance or resistance to some compounds in other cotton growing regions, but resistance to insecticides has not become a major problem in the Western Region. Although pyrethorids are effective for pink bollworm, it is best to save them for damaging populations of tobacco budworm or bollworm. Using pyrethroids unnecessarily for pink bollworm could increase the level of resistance in *Heliothis* and other secondary pests.

The treatment threshold in Arizona and southern California is 10% infested bolls. In New Mexico, the threshold is 5% to 10%. These thresholds are useful only when a significant number of susceptible bolls is present.

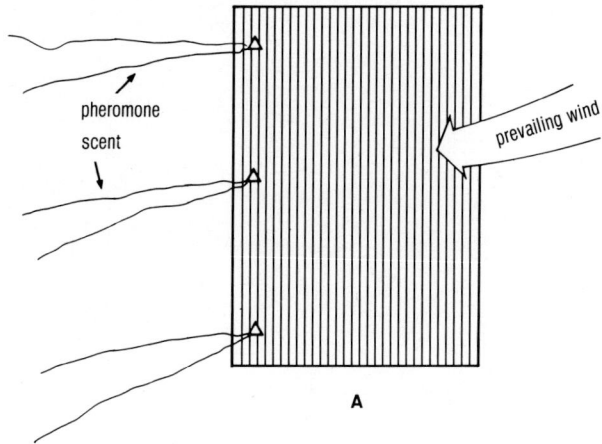

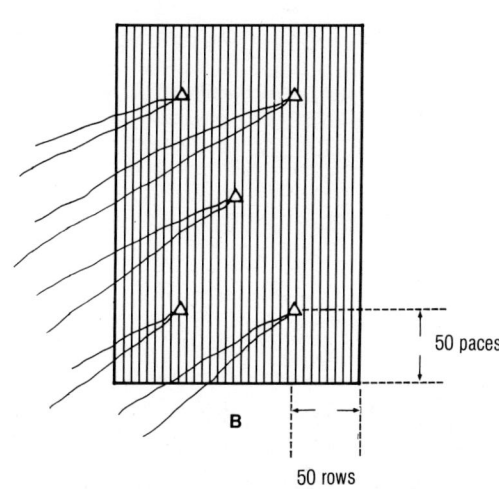

Figure 30. Pheromone traps placed at the edge of a field (A) produce misleading results when wind blows the scent into unplanted areas. Traps placed at least 50 paces and 50 rows into the field (B) are more reliable because the scent is kept mostly within the planted area.

Applying them too early in the season or during a cutout period, when few green bolls are developing, will result in unnecessary treatments. In Arizona, a series of three or four treatments 6 days apart is recommended once the percentage of infested bolls exceeds the threshold. In southern California and New Mexico, a single treatment is recommended, to be repeated only if a new set of boll samples a week later shows that the infestation is still above the threshold.

A special method of applying insecticides for pink bollworm control, the "attracticide" method, has been widely introduced in the low deserts. This method uses tiny plastic fibers or flakes impregnated with pheromone as a bait. The bait is sprayed onto foliage in a sticky mixture that also contains a relatively low rate of insecticide,

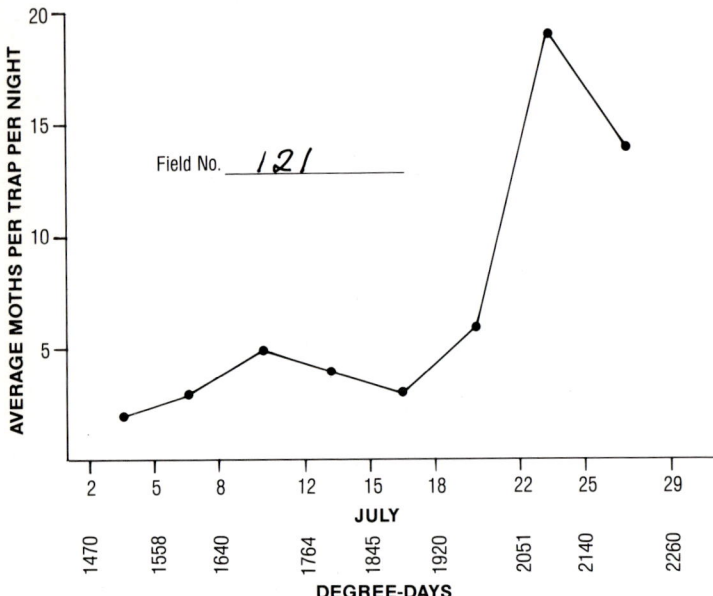

Field No. *121*

Figure 31. When recording pheromone trap results for pink bollworm, plot the average number of moths per trap *per night* according to date and/or degree-days. Put each data point midway between two check dates. If you plot trap results exactly on the date when you check the traps, it will imply that all the moths were caught on the last night before the traps were emptied, and the position of peaks on the graph will be misleading.

Exit holes left by pink bollworms are about ⅛ inch (2 to 3 mm) in diameter.

Bolls damaged by pink bollworm may fail to open and often rot.

The delta trap's sticky inner surface holds males attracted by the pheromone lure. While plants are small, keep traps within a foot of the ground; move them up as plants grow.

usually a pyrethroid. Male moths attracted to the fibers or flakes then contact the insecticide and may be killed.

Mating disruption, which involves the broadcast application of pheromone to a whole field, is intended to suppress reproduction by preventing males from finding females for mating. Although it has apparently reduced some early season populations, mating disruption is still experimental. Consult farm advisors for recommendations on using this technique.

Mass trapping—removing males from the population by trapping them—is another way to interfere with mating and reproduction. This method has been tested using pheromone-baited oil traps in Pima cotton in Arizona. Eight to 10 oil traps per acre are needed for best results. Set them out before squaring, or, at the latest, when the first squares are 10 days old. Attach traps to stakes in the rows and leave them in place; you need not move the traps up as in monitoring.

Other Caterpillars That Injure Fruit

Aside from those already discussed, the most common caterpillar that chews into squares and bolls is the beet armyworm (page 76); it can produce injury similar to that caused by bollworms and budworms. Although it occasionally causes significant damage to fruit, it feeds mostly on foliage. Other foliage-feeding caterpillars, when present in very large numbers, may also injure squares or bolls, but their injury to fruit is rarely significant.

The cotton square borer, *Strymon melinus*, is a slug-shaped, green caterpillar covered with short, dense hairs that give it a velvety appearance. The larva is up to ½ inch (1 cm) long; it can retract the head into the thorax so that it is not visible from above. Square borers feed on leaves when small but later bore into squares or bolls. Infestations are rare because most larvae are killed by parasites, but isolated larvae are found occasionally. The adult is a small gray butterfly with short tails on the hind wings.

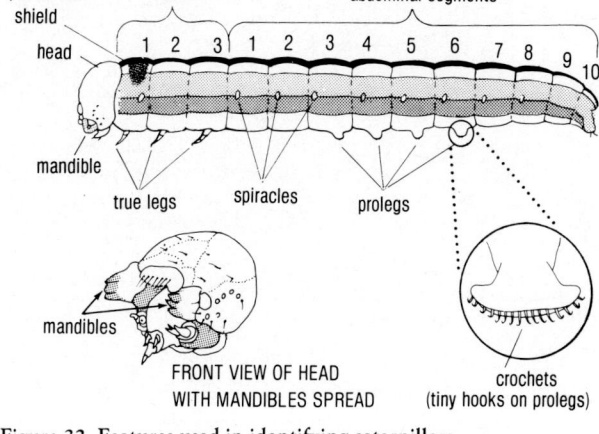

Figure 32. Features used in identifying caterpillars.

Key to Caterpillars and Similar Larvae on Cotton in the Western Region

The purpose of this key is to distinguish important pest caterpillars from other species commonly found on cotton in the Western Region. Larvae of the boll weevil and leafmining flies are included because they can be confused with caterpillars. To simplify the key, several species found only occasionally have been omitted; ask a farm advisor or county agent for help in identifying specimens that do not fit the key.

To use the key, start at the first pair of descriptions and choose the one that applies to your specimen, then continue to the number indicated. Read *both* descriptions and compare the specimen with the drawings before proceeding. When you arrive at a name, compare the specimen with the photos and descriptions in this chapter. If possible, take several specimens through the key individually; collections of larvae often include more than one species.

Most features mentioned in the key can be seen with a good hand lens, but you may need a low power microscope to see some features on small larvae. The key works best for specimens in the third instar or larger.

1. Larvae up to 1/8 inch (3 mm) long, feeding in mines between leaf surfaces
   See 17

   Larvae feeding in various sites on plant, but never in leaf mines; size variable
   See 2

2. True legs present on thorax and at least two pairs of prolegs on abdomen (Figure 32)
   See 3

   True legs and prolegs absent; larvae up to 1/4 inch (6 mm) long, curled into C shape, feeding inside squares or bolls
   Boll Weevil

3. Larvae with numerous long, hairlike bristles arising in bunches from several tubercles on each segment
   Saltmarsh Caterpillar

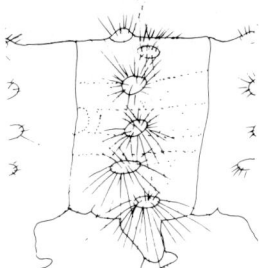

typical abdominal segment

Larvae with only sparse bristles, or with numerous short spines not arranged in bunches
   See 4

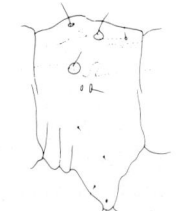

typical abdominal segment

4. Horn present on eighth abdominal segment

    Whitelined Sphinx

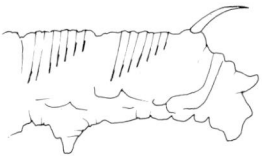

Horn absent

    See 5

5. Larvae sluglike in shape. Head usually retracted into thorax, not visible from above. Skin velvety with numerous short bristles. Rarely more than a few larvae per field

    Cotton Square Borer

Not sluglike; head always visible from above

    See 6

6. Larvae boring into tips of stems or feeding in shelters constructed of leaves or bracts webbed together with silk. Feeding singly, not in groups on one leaf; rarely found inside squares or bolls

    See 7

Larvae exposed on foliage or feeding inside squares, flowers, or bolls; never in stems or shelters. (Beet armyworm may spin silk threads over a leaf in early instars, but does not web leaves together.)

    See 8

7. Anal comb present. Three fine bristles (setae) arranged in a group in front of spiracle on first thoracic segment. Early instars may bore in stems. Up to 1/2 inch (12 mm) long

    Omnivorous Leafroller

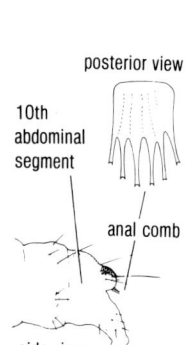

posterior view

10th abdominal segment

anal comb

side view

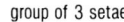
group of 3 setae

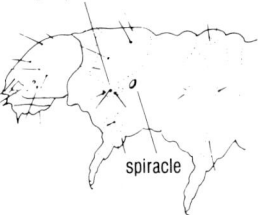

spiracle

Anal comb absent. Only two setae in front of spiracle (check both sides of specimen). Up to 1 inch (25 mm) long

    Webworms and Leaftiers

8. Prolegs long and slender. Body tightly constricted between abdominal segments; each segment with two black and four white dots on top. Larvae up to 3/8 inch (9 mm) long, skeletonizing leaves, dropping on silk thread when disturbed. Desert valleys only

    Cotton Leafperforator

Prolegs short, as in Figure 32. Appearance and behavior not as described

    See 9

9. Well developed prolegs present on segments 3 through 6 of abdomen (Figure 32)

    See 10

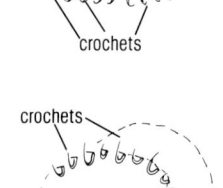
prolegs

Prolegs lacking or greatly reduced on abdominal segments 3 and 4

    Cabbage and Alfalfa Loopers

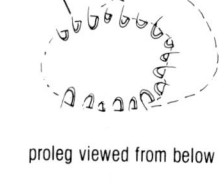
prolegs

10. Crochets on prolegs of abdominal segments 3 through 6 arranged in a single line along inner edge of proleg. Larvae up to 2 inches (5 cm) long

    See 11

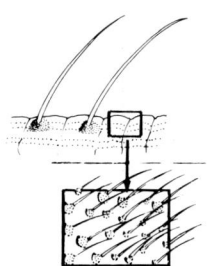
crochets

Crochets on prolegs of segments 3 through 6 arranged in nearly circular pattern. Larvae up to 3/8 inch (9 mm) long, pink when mature, feeding in squares, flowers, or bolls; rarely exposed on foliage

    Pink Bollworm

crochets

proleg viewed from below

11. Much of body surface covered with short, whiskerlike spines in addition to several longer bristles on each segment. Spines visible at 10x to 20x magnification

    See 12

Skin smooth or with blunt bumps when seen at 10x to 20x; short spines lacking

    See 13

12. Short spines present on tubercles on top of eighth abdominal segment. Large tooth (retinaculum) usually present on inner side of mandible, but may be worn or broken in some specimens
Tobacco Budworm

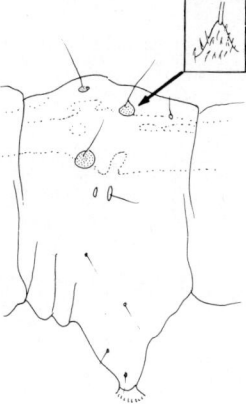

8th abdominal segment

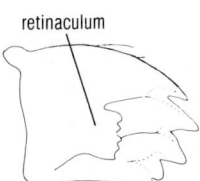

retinaculum

inner surface of mandible

Spines absent on tubercles on eighth abdominal segment; retinaculum absent
Bollworm

Note: It is not possible to distinguish tobacco budworm from bollworm reliably before the third instar

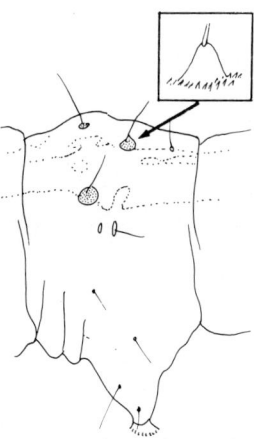

8th abdominal segment

13. Skin appears bumpy or granular at 10x to 20x magnification
Cutworms, various species

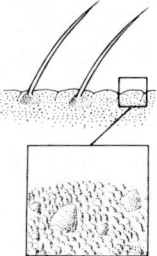

Skin appears smooth at 10x to 20x
See 14

14. Centers and borders of spiracles uniformly dark in color, usually black
Cutworms, various species

Centers of spiracles white or brown, contrasting with darker borders
See 15

15. Bristles on top of abdominal segments arising from tubercles distinct from surrounding skin
Fall Armyworm

Bristles on abdominal segments not arising from distinct tubercles
See 16

16. Centers of spiracles white; black spot usually present on second segment of thorax
Beet Armyworm

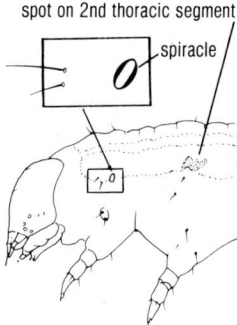

spot on 2nd thoracic segment

spiracle

Centers of spiracles brown; black spot present on first abdominal segment, just above spiracle
Yellowstriped and Western Yellowstriped Armyworms

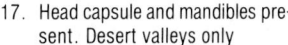

spot above spiracle

17. Head capsule and mandibles present. Desert valleys only
Cotton Leafperforator, early instars

Head capsule and mandibles absent   **Leafminers**

Boll Weevil
Anthonomus grandis

The boll weevil is a major cotton pest in Mexico and the southern U.S. Adults and larvae feed mainly in squares, but they also attack bolls. Recent outbreaks in central and western Arizona were originally associated with stub cotton, but the weevil persisted as a damaging pest of planted cotton even after stub cotton was eliminated. Boll weevils have also been found in southern California.

It is uncertain whether it may be possible to eliminate or contain the boll weevil in the areas now infested. Its widespread, permanent establishment could have a major impact on western cotton, as infested fields require insecticide applications that can trigger outbreaks of such secondary pests as tobacco budworm. The key management strategy is to reduce overwintering populations by early harvest and prompt shredding and plowdown.

Description

Adult boll weevils are gray or brown and have a snout about half as long as the rest of the body. Mouthparts are at the tip of the snout, which also bears the elbowed antennae. You can distinguish boll weevils from other weevils found on cotton by the pair of spurs on each front leg (Figure 33).

The elongate, white eggs are inserted in characteristic punctures in squares and bolls. The larvae are legless, cream colored grubs with light brown heads; they are usually curled into a C shape. The absence of legs distinguishes them from all other larvae commonly found inside squares or bolls. The pupa has the general form of the adult, including the snout, except that the appendages are held tightly against the body.

Seasonal Development

In Arizona, boll weevils continue development all year as long as bolls or squares are available. There may be as many as seven or eight generations a year. The weevils complete a generation in as little as 3 weeks in summer; the overwintering generation may require 2 months or more.

In spring before squaring, adults feed on terminal buds and petioles of cotton and also on the pollen of wild cotton family plants such as globe mallow, *Sphaeralcea* spp. They can fly several miles in search of host plants. When a male begins feeding, it releases a pheromone that attracts both sexes from a considerable distance. As new males arrive, they also produce the scent, attracting yet others. For this reason, spring populations are usually clumped in small parts of a field and are easy to miss in monitoring. Infestations may not be apparent until midsummer, after two or three generations have spread over a much larger area.

When squares are available, adults chew into them with their snouts, consuming pollen, stamens, and pistils. They will also feed on the immature lint of green bolls.

Egg laying begins when first squares are one-third to one-half grown. Eggs are placed in punctures similar to feeding punctures, and the larvae feed on the internal structures. Infested squares usually flare and drop, and larvae often complete development in squares on the ground. Although they prefer squares for both feeding and egg laying, the weevils will also attack bolls, where the larvae feed mostly on the developing seeds. Egg punctures are usually sealed with a plug of feces. As they feed, the larvae create a chamber of frass and feces where they later pupate. There is usually only one larva in each square or boll, but there may be more if the population is high compared with the number of fruiting structures.

The larval period is about 7 to 10 days in summer. After pupation, which takes about a week, adults of summer generations usually chew their way out of the square or boll and feed on pollen for 1 to 4 days before they begin mating and laying eggs. However, adults that develop in mature bolls late in the season may remain there for 4 or 5 months, emerging only when the bolls are broken in shredding or softened by moisture. These adults are a major part of the overwintering population in fields not plowed down before spring. Other adults leave cotton fields in the fall. Some may pass the winter in diapause, but under Arizona conditions at least part of the population remains active through the winter. Adults have been caught in pheromone traps every month of the year. Mild winters that favor stubbing also favor the boll weevil; they allow survival of more overwintering adults.

A native race of the boll weevil, known as the thurberia weevil, feeds on wild cotton, *Gossypium thurberi*, in southern Arizona. Minor late-season infestations of thurberia weevil are common in fields close to desert mountain ranges where wild cotton grows. However, most weevils infesting commercial cotton in the desert valleys are more similar in habits to the boll weevil of the Southeast even though they are identical in appearance to the thurberia weevil. Thought to have originated in Mexico, they are known as the Mexican or Sonoran race of the boll weevil.

Damage

Squares punctured by adults usually flare and drop. You can identify injured squares by the punctures and the yellow frass usually found on the outside. Flowers that

Figure 33. The boll weevil has two spines on each front leg.

Adult boll weevils commonly feed in flowers during the day.

The boll weevil larva is legless and usually curls into a "C" shape. It forms a chamber of frass as it feeds among the seeds.

Squares injured by adult boll weevils have round punctures and bright yellow frass.

Punctures where boll weevils have laid eggs are usually sealed with a plug.

The boll weevil pupa is inside the chamber created by the larva.

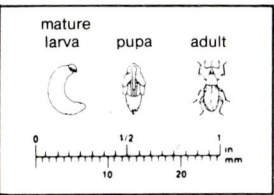

BOLL WEEVIL

open from injured squares have small, round holes in the petals and some of the anthers are damaged.

Infested bolls may either drop or remain on the plant. They fail to open and fluff normally. Damaged locks have a distinctive feeding and pupation chamber, fleshy orange at first and later dark colored and tough. Some or all of the seeds are eaten and most of the lint is discolored. Exit holes in bolls resemble those left by pink bollworm, but the pupation chamber serves to distinguish weevil-damaged bolls.

Management

As in managing pink bollworm, managing boll weevils requires keeping the overwintering population low by harvesting as early as possible, then shredding all crop debris and discing or plowing it under to a depth of 6 inches (15 cm). Weevils may survive the shredding, but they cannot emerge if they are buried. Early harvest limits the number that escape to pass the winter outside the cotton field. A single field not plowed down on schedule may serve as a source of weevils within a radius of several miles. As part of plowdown and sanitation, eliminate volunteer cotton from roadsides, ditchbanks, and fields of other crops. In some parts of the U.S. cotton belt, shifting to short-season cotton has kept weevil populations to manageable levels.

Arizona is the only western state where specific monitoring and treatment practices are currently recommended for boll weevil. In California, state and county agencies have begun a program to eradicate it, but there is no recommendation for crop managers other than to remain alert for possible infestations. However, most techniques recommended for Arizona could also be used in southern California if necessary.

Monitoring. Monitoring methods for boll weevil vary according to the stage of crop growth. Early in the season, pheromone traps are useful for detecting adult weevils. Traps baited with the boll weevil pheromone, grandlure, should be set out as early as possible—preferably at the first of the year or at least by planting time. Later, once the weevils have begun feeding on cotton, pheromone produced by the weevils overpowers the lure in traps, so trap catches may drop to zero even though an infestation is increasing. Follow the manufacturer's directions for handling and placement of traps.

Several other species of weevils may appear in boll weevil traps. Some, such as alfalfa weevils, are easy to distinguish from boll weevils since they have short, broad snouts and lack the spurs on the front legs. Others, however, could easily be confused. *Anthonomus peninsularis*, a species that feeds on wild globe mallow and is not a pest, is closely related to the boll weevil; it has the same general appearance, including the spurs on the front legs. Contact your local Extension agent for help in correct identification.

Between crop emergence and early squaring, adult boll weevils may begin congregating on the plants. Infestations are hard to find at this stage, but you can look for them by examining plants in randomly chosen sections of row. Look carefully for adults on plant terminals.

As soon as plants have an average of about three 10-day-old squares, pick samples of squares to look for punctures. This is the main monitoring technique for most of the season. In each field, pick at least 100 ten-day-old squares at random once a week or more. Sample squares should be about the size of a pencil eraser. Treatment thresholds in the southern states are often based on the percentage of small squares punctured by weevils. Similar thresholds may be established in the Western Region if weevil infestations persist. While sampling squares, be alert for adults in flowers; they are easy to spot when feeding on pollen in open flowers.

Control. Once a field is infested with boll weevils, repeated insecticide applications are needed for control. Because immature stages are protected in squares or bolls, insecticides must be aimed at adults during the day, when they are feeding and laying eggs in the tops of plants. Daytime applications of materials commonly used for weevil control present a major hazard to bees.

Arizona's current control program is based on the premise that it may still be possible to reduce the weevil to its former status as a minor, occasional pest. If this proves impossible, control recommendations will be modified. Because experience with boll weevil under western conditions is limited, new management practices may be developed as new information becomes available. Keep in touch with farm advisors and county agents concerning recommendations.

Early season treatments are recommended in all Arizona fields where weevils were found the previous season. Apply the first treatment when the first squares are 10 days old. Follow with one or two more treatments at 5-day intervals. After the early season treatment, treat again whenever you find adult weevils or freshly punctured squares. In heavily infested fields, treatment may be needed every 4 to 6 days throughout the season.

In fields where the weevil has not been found before, set out pheromone traps and follow sampling procedures previously outlined. Begin treatments if you find one or more weevils in the traps, or if you find weevils or punctured squares on plants.

If you find weevils at any time, a late-season cleanup treatment is recommended. Apply a suitable insecticide along with the defoliant; then treat once or twice before harvest to kill adults that may emerge from infested squares or bolls present at the time of defoliation.

If you find boll weevils in pheromone traps during winter or if you know from the previous year that they are present in your area, it may be worthwhile to create a trap crop by planting a few rows of cotton 2 or 3 weeks early. If the early rows begin squaring before the rest of the field, the weevils will be concentrated there. Treating weevils in trap rows early in the season may delay the need for insecticides in the rest of the field. Another suggestion for creating a trap crop is to spray a few rows of cotton with the boll weevil pheromone. This method has not been tested adequately under western conditions, and it is not likely to work once squaring has begun.

Parasitic wasps may attack a proportion of the boll weevil larvae late in the season, but they have little impact on populations.

Lygus Bugs
Lygus hesperus and others

Lygus bugs, common everywhere in the western U.S., are important pests of cotton in the San Joaquin Valley and in New Mexico, at least in occasional years when populations are high. In the desert valleys, lygus bugs are usually controlled by insecticides applied for pink bollworm, but they still occur often enough that regular monitoring is recommended. Lygus injury to cotton is due mostly to feeding on small squares, although the bugs may also injure young bolls. Infestations begin with adults that fly in from weeds or other crops, particularly alfalfa. Treatment decisions for lygus bugs in cotton are based on sampling with the sweep net.

The four species of lygus bugs that occur on cotton in the western U.S. are all similar in appearance and habits, and it is not necessary to separate them in monitoring. *Lygus hesperus*, the most common species, occurs throughout the region; it is the only lygus bug that regularly damages San Joaquin Valley cotton. *L. elisus* sometimes invades San Joaquin Valley cotton in large numbers, but infestations usually last only a few days. The less common *L. desertinus* occurs in Arizona, southern California, and New Mexico. The tarnished plant bug, *L. lineolaris*, is a major pest elsewhere in the cotton belt, but it is uncommon in the West.

Description

Adults of all species are green, straw yellow, or brown with a conspicuous yellow or pale green triangle in the center of the back. The antennae are long, slender, and usually reddish or brown. The light green first and second instar nymphs are sometimes confused with aphids because of their small size; however, lygus nymphs move much faster than aphids, and they have a distinctive reddish color on the tips of the antennae. Older nymphs have a characteristic five black dots on the back: two dots on the first segment of the thorax just behind the head; two more on the next segment; and one spot in the center of the abdomen. This pattern distinguishes older *Lygus* nymphs from all other nymphs commonly found on cotton.

The bug most likely to be confused with lygus bugs is the cotton fleahopper. Adult fleahoppers are only about half as large as adult lygus bugs. Fleahoppers of all stages are uniformly pale green with small black specks all over the body.

The narrow, cylindrical eggs are inserted in plant tissue, mostly in the petiole, where it joins the leaf blade. Eggs may also be laid in stems and squares. When lygus bugs are extremely numerous, the surface of petioles may be roughened by large numbers of eggs.

Seasonal Development

Lygus bugs have several generations a year, but usually not more than three develop on cotton. Infestations begin when adults fly in from winter and spring hosts that have been harvested or have begun to mature and dry out. Alfalfa, a preferred host that harbors the bugs all year, is often the main source of infestation. When alfalfa is cut, lygus bugs fly to other nearby hosts including cotton.

Many common broadleaf weeds are lygus hosts, including redroot pigweed, lambsquarters, and related plants, as well as knotweed, wild sunflower, and others. Mustard family annuals such as shepherdspurse, London rocket, and black mustard are important spring hosts. Lygus bugs often migrate to cotton when weeds are mowed or disced in orchards and vineyards or when weedy fields of such annual crops as sugarbeets or tomatoes are harvested. Weeds in roadsides, ditchbanks, and other waste areas are also important sources of lygus bugs; adults often fly to cotton when the weeds start to dry out in spring or summer. Lygus bugs may also migrate to crops from such rangeland weeds as Russian thistle, lupine, and tarweed. This is especially common near the foothills on the San Joaquin Valley's western and southern edges. Late spring rains that delay the drying of weed hosts may allow lygus bugs to complete an extra generation and may increase the number that later move into cotton.

Lygus bugs that fly into cotton sometimes leave again in as little as a day. Their movements are affected by the crop's condition, the stage of fruiting development, and

First instar lygus bug (left) is about the same size as a cotton aphid, but can run rapidly and has the tips of the antennae reddish brown.

Lygus nymphs may be green or brown. Nymphs in the third instar or older have a pattern of five black dots on the back.

Anaphes ovijentatus, a parasitic wasp, lays its egg in the egg of a lygus bug.

When lygus bugs are causing squares to shed, you may see injured squares in the sweep net.

Shriveled, dried out squares injured by lygus bugs are often trapped in young foliage. Note the scar that remains where the square hs been lost.

Lygus bugs vary in color, but always have a prominent triangle in the center of the back. This is *Lygus hesperus*, the most common lygus species in the Western Region.

the availability of other hosts. Plants stressed for water are less attractive than normally, while rank plants are more attractive. Lygus bugs that remain in cotton only a short time do not cause significant loss unless the migrating population is extremely large; even then, damage is usually limited to the edge of the field closest to the previous host. However, lygus populations that remain long enough to begin reproducing can be very destructive. Young nymphs feed on tender vegetative tissues, but in the third or fourth instar the bugs begin to attack small squares.

Damage

Losses are due mostly to the destruction of squares less than 1/5 inch (5 mm) long. Lygus bugs pierce the squares and consume anthers and other tissues; the squares then shrivel, turn brown, and drop from the plant in a day or so. Larger injured squares may remain on the plant if the damage is not too extensive; otherwise, the bracts flare before the square drops. Flowers that develop from injured squares have some black and shriveled anthers, and they may have wrinkled, distorted petals. Bolls from injured squares may be lopsided because of incomplete pollination.

Removal of squares reduces the demand placed on the plant's energy resources by fruit development. With reduced demand, more energy is available for vegetative growth. When a large proportion of the small squares is destroyed over a long enough period early in the season, plants may become tall and spindly and remain unfruitful.

Injury due to caterpillars, soft rot, or water stress may also cause small squares to shrivel and drop. To distinguish these injuries from lygus damage, examine the squares with a hand lens. In squares injured by lygus bugs, the anthers are brown and shriveled and the pistil may be missing, but the hole left by the slender mouthparts is too small to be seen. Squares damaged by caterpillars, such as bollworms, show signs of chewing and often have a definite entry hole, usually at the base. Squares may also be destroyed by a brown, soft rot that occurs when thrips or other insects introduce decay-causing bacteria; this is usually a minor factor and occurs only on the earliest squares. Squares shed due to water stress or late-season thinning simply dry up. They have no internal parts missing, have no signs of chewing, and are not rotten inside.

On small bolls, lygus bugs pierce the boll wall to feed on the developing seeds, which may then fail to develop.

Feces of lygus bugs contain a substance that causes black, sunken spots on the surface of young bolls.

Young bolls injured by lygus bugs have shrunken and stained seeds. A normal boll is at right.

When lygus bugs injure the growing terminal of the main stem early in the season, plants develop an abnormal branching pattern.

LYGUS BUGS

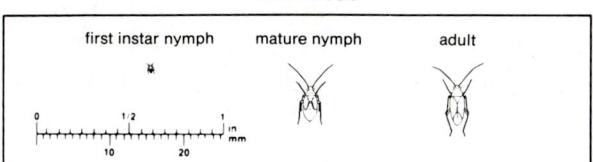

first instar nymph   mature nymph   adult

Lint around injured seeds is stained yellow, may not mature normally, and may be lost in harvesting. Bolls are susceptible to lygus injury until about 10 days after flowering, when their walls become too tough to pierce. Injured bolls usually have sunken, black or brown spots on the outer surface caused by a toxin in the bugs' feces. Injury to bolls occurs mostly in the latter part of the season, when the number of squares declines. Boll damage has much less effect on yield than does earlier damage to squares.

Lygus injury to terminal buds early in the season can prevent normal growth of the main stem and may result in abnormal vegetative branching—the "candelabra effect." Affected plants yield much less than normal plants, although the overall impact of this kind of damage is minor compared with damage caused by feeding on squares.

Management

Because lygus bugs migrate to cotton from other hosts, management begins with assessing their populations outside the field. Check for them on weeds, in nearby alfalfa, and in other crops, and keep in touch with your Extension agent or farm advisor for area-wide information on lygus populations. Proper management of alfalfa harvest can reduce damaging migrations to cotton. The need for insecticides in cotton must be evaluated carefully on a field by field basis, as treatments may result in secondary outbreaks of spider mites or other pests.

Biological Control. The most important natural enemies of lygus bugs are bigeyed bugs, which attack eggs and young nymphs. Other predators, such as damsel bugs and collops beetles, also feed on the eggs and nymphs, and crab spiders attack lygus adults. A parasitic wasp, *Anaphes ovijentatus*, attacks the eggs of lygus and other plant bugs throughout the western U.S., and another, *Leiophron uniformis*, attacks young nymphs in the deserts. Predators and parasites can help to keep lygus populations low. However, these enemies are not likely to be active in fields treated with broad-spectrum insecticides and are seldom abundant enough to prevent damage when large numbers of lygus adults migrate into cotton.

Management of Lygus Bugs in Alfalfa Hay. Vigorously growing alfalfa attracts lygus bugs more than does cotton. Although lygus bugs are important pests of seed alfalfa, they cause little or no injury to alfalfa hay. Where alfalfa hay is grown, you can prevent damaging migrations by leaving an uncut strip or check at each cutting. At the next cutting, harvest that strip or check, but leave another uncut, so there is always a stand of alfalfa to attract the lygus bugs. Another approach is to divide large hay fields in two; then stagger irrigation and harvest schedules. Where several small alfalfa fields are close together, stag-

ger the harvest so that all the fields are not cut in the same week. In addition to its effect on lygus bugs, strip cutting helps to preserve natural enemies in the alfalfa. Another way to use alfalfa is to interplant strips of it every 300 to 500 feet (90 to 150 m) in the cotton field. The strips must be adequately watered so that the alfalfa will remain attractive to lygus bugs.

Management of Lygus Bugs in Safflower. Safflower can be a major spring host for lygus bugs that later move to cotton. Until recently, routine treatment of safflower fields was recommended in the San Joaquin Valley to reduce lygus populations in cotton. In the last few years, however, populations in safflower have been so low that treatments were not needed. The reason for the change is not known, but some researchers relate it to the introduction of new safflower varieties; more study is needed to determine whether there has been a significant change in the status of safflower as a lygus host.

The method previously recommended for scheduling insecticide treatment of safflower in the San Joaquin Valley is based on the accumulation of degree-days above 52° F. The objective of this method is to schedule a single treatment at the time when most lygus nymphs are in the third to fifth instar. At this point in the population cycle, there are few winged adults able to migrate and there is too little time before the safflower harvest for another generation to develop. Farm advisors in areas where safflower is grown will have available information on preventive treatments for lygus bugs.

Treatment for lygus bugs in safflower is still recommended in Arizona. Apply a single treatment after the final irrigation, but before the crop begins to dry down. There is no degree-day system for use in treating Arizona safflower.

Weed Control. Increased use of herbicides in many crops has in recent years reduced the potential for lygus bug migrations. However, there is room for improving control in roadsides, fencerows, and waste areas. The best strategy is to destroy weeds before cotton plants emerge by using tillage or appropriate herbicides. If you destroy host weeds when cotton is already growing, the bugs may then fly into the cotton. Leaving a stand of weeds can have the same effect, since the bugs will migrate once the weeds start to dry out.

Monitoring in the San Joaquin Valley. The critical period for lygus injury in the San Joaquin Valley is from the start of the third week of squaring to the end of the sixth week, or from about 700 to 1400°D (base 60° F) after planting. This is when excessive loss of small squares has the greatest effect on yield. During this period, the treatment threshold is not fixed. Instead, you determine it each week—a total of three times—by counting squares in sample sections of row. When the number of squares is high, the threshold is also high, as it would take a large number of bugs to damage enough squares to affect yield; when the number is low, the threshold is proportionally lower. There is no treatment threshold in the first 2 weeks of squaring, although you should watch for unusually large infestations. After the critical period, most normal fields can tolerate a fairly high number of bugs without loss.

Treatment thresholds are expressed in terms of the number of bugs found in 50 sweeps with a standard net. Follow directions below for taking the samples. Always use the average of at least four sweep net samples in each field to determine whether the number of bugs exceeds the threshold. Take a set of sweep net samples weekly. If the number of bugs approaches or exceeds the threshold, return in 3 or 4 days and sample again. Treat only when the average exceeds the threshold on two successive sample dates. Because lygus bugs may suddenly migrate out of cotton, treating on the basis of a single sample date will result in too many unnecessary treatments.

Counting Squares. Each week during the critical period, count squares in four places in each field. First, visually divide the field into four equal areas. In each area, pick a spot at random where plant growth is typical of the whole area and count all the squares on all the plants in a section of row 40 inches long. Add the results of the four samples; then multiply the total by .03 to find the week's treatment threshold. For example, if you counted 104, 96, 88, and 112 squares in a series of four samples for a total of 400 squares, the threshold for that week in that field would be 400 x .03, or 12 bugs per 50 sweeps. Treatment would be needed if you found an average of 12 or more lygus bugs per 50 sweeps on two successive checks 3 or 4 days apart.

Sweep Samples. Take a series of sweep net samples weekly in each field. Always use a standard net with a diameter of 15 inches. Take one sample in each quarter of the field in fields up to 80 acres. Take more samples in larger fields.

Each sample consists of 50 sweeps across a single row of cotton. Walk briskly down the row and swing the net in front of you so that the lower edge of the rim strikes the plants about 10 inches (25 cm) below the top. Keep the lower edge slightly ahead of the upper edge (Figure 34). Keep the sweeps far enough apart that you do not sweep plants that have already been jostled by the net. If the sweeps are too close together, lygus bugs may fly or drop from the plants and will be missed. Keep the net moving fast enough that adults cannot fly out.

After each set of 50 sweeps, count all the lygus bugs in the net, including nymphs, and write the number on your monitoring form (Figure 35). Be careful not to confuse aphids with small nymphs. When you have finished

Figure 34. Use a standard 15-inch sweep net to sample for lygus bugs. Swing the net so that the lower rim strikes plants about 10 inches below the tops. The lower edge of the rim should precede the upper edge slightly.

sampling the field, calculate the average number of bugs per 50 sweeps.

Don't take sweep samples when the wind is stronger than 15 mph; lygus bugs move down into the canopy when it is windy and will be missed by the net. You can sample at any time of day, although it is difficult to count small nymphs if you sweep while plants are wet with dew.

Monitoring Before and After the Critical Period. There is no specific monitoring procedure and no treatment threshold for lygus bugs before the third week of squaring. However, it is important to be alert for unusually large populations and to make sure the crop is developing normally during the first 2 weeks of squaring. Even before planting, check for lygus bugs on weeds or in adjacent crops. Keep in touch with farm advisors who may have information on lygus populations on other crops or on rangelands. These populations serve as a warning of possible migrations to cotton later.

LYGUS BUG MONITORING
SAN JOAQUIN VALLEY
THIRD THROUGH SIXTH WEEKS OF SQUARING ONLY

STEP 1: count squares once each week.

In each field quadrant, count all squares in a 40-inch section of row chosen at random.

QUADRANT	SQUARES
1	____
2	____
3	____
4	____
total:	____

Divide the square total by 100, then multiply by 3:

$$\frac{\underline{\hspace{1cm}}}{100} = \underline{\hspace{1cm}} \times 3 = \boxed{}$$

The number in the box is the treatment threshold for this week.

STEP 2: Take at least one sweep sample in each quadrant. One sample is 50 sweeps across one row with a 15-inch net. Count all lygus bugs, including nymphs, in each sample:

SAMPLE NO.	LYGUS BUGS
1	____
2	____
3	____
4	____
5	____
6	____
7	____
8	____
9	____
10	____
total:	____

Divide the total by the number of sweep samples to find the average bugs per 50 sweeps:

$$\frac{\text{(total)}}{\text{(samples)}} = \underline{\hspace{2cm}} \quad \text{(average)}$$

Treat when the average exceeds the treatment threshold on *two consecutive* sample dates 2 or 3 days apart.

Figure 35. Sample form for monitoring lygus bugs in the San Joaquin Valley.

After the sixth week of squaring, you no longer need to count squares, but you should continue weekly sweep samples until about 2 weeks after peak squaring. Where the crop has been squaring and setting bolls normally, the treatment threshold is 20 to 30 bugs per 50 sweeps. A lower threshold may be needed in replanted fields and in fields that have failed to set fruit earlier.

Monitoring in the Desert Valleys. The treatment threshold for lygus bugs in desert areas is 10 bugs per 50 sweeps. Sweep net sampling is conducted in the same way as in the San Joaquin Valley. Sample once a week, taking at least four sweep samples in each field, starting in the third week of squaring. If the population exceeds the threshold, sample again in 3 or 4 days, and treat if the population is still above the threshold on the second sample date. In fields that are managed for a top crop, you may need to monitor during squaring in the second fruiting cycle. However, late season lygus injury is uncommon in fields treated repeatedly for pink bollworm, tobacco budworm, or boll weevil, because insecticides used for these pests usually kill lygus bugs.

Monitoring in New Mexico. Monitoring lygus bugs in New Mexico is similar to that in the San Joaquin Valley. Because of the shorter growing season, however, excessive injury to squares any time during squaring can reduce yield, so tolerance for the bugs is somewhat lower. This difference is reflected in the formulas for determining the treatment thresholds.

Squaring in New Mexico lasts about 4 weeks. Each week, count the squares on all plants in four 40-inch samples of row in each field. Count one sample in each quarter of the field and add the results. In central and western New Mexico, multiply the total by .025 to get the week's treatment threshold. For example, if you find a total of 380 squares, the threshold should be 380 x .025, or 9.5 lygus bugs per 50 sweeps. In eastern New Mexico, multiply the number of squares by .02. For the same total of 380 squares, the threshold should be 380 x .02, or 7.6 bugs per 50 sweeps.

Follow directions for taking sweep net samples in the San Joaquin Valley. Treat when the average number of bugs in four 50-sweep samples exceeds the threshold on two successive sample dates 2 or 3 days apart.

Insecticides for Lygus Bugs. Lygus bugs are resistant or tolerant to several common insecticides (Table 8). Some materials may have adverse effects on natural enemies or on the crop. Don't apply an insecticide unless monitoring shows the population has exceeded the treatment threshold.

Resurgence of lygus populations is common. No available insecticides will kill lygus eggs, but most will destroy natural enemies, so nymphs that hatch from eggs present at the time of application will often develop with little interference from natural enemies. This problem is especially noticeable with organophosphates. Timing of the resurgence depends on the insecticide's residual properties—the longer the residue lasts, the later and the more limited the resurgence.

Treatment for lygus bugs may also cause outbreaks of secondary pests, especially spider mites. There is essentially no natural control of spider mites after a lygus treatment, so any mite population present at the time of application is likely to increase rapidly. Insecticides that often increase mite populations include acephate (Orthene), methyl parathion, and permethrin (Ambush/Pounce). Insecticides applied after peak bloom may contribute to outbreaks of bollworm, tobacco budworm, beet armyworm, and loopers.

Other Plant Bugs

Several kinds of plant bugs, other than lygus bugs, are found on western cotton. In most areas, these species occur only in small numbers and have little effect on the crop. Some plant bugs feed largely on other insects.

The cotton fleahopper, *Pseudatomoscelis seriatus*, occurs throughout the Western Region, but is common on cotton only in eastern New Mexico. Fleahoppers are about half as large as lygus bugs. Adults and older nymphs are uniformly pale green with tiny black specks all over the body; nymphs never have reddish antennae or the pattern of dots found on lygus nymphs.

Fleahopper infestations are most common in fields near uncultivated land with abundant weed hosts. The main weed host is *Croton texensis*, often called goatweed or doveweed. Fleahoppers may migrate to cotton when weed hosts mature in spring or summer. Damage to cotton is similar to lygus damage; impact on the crop is greatest in early squaring.

Monitoring techniques for fleahoppers in New Mexico are the same as for lygus bugs—count squares each week to determine the treatment threshold, and take sweep net samples once or twice a week. Treat when the average number of fleahoppers in 50-sweep samples exceeds the threshold on two successive dates 2 or 3 days apart. Use this formula to determine weekly thresholds:

Nymphs and adults of the cotton fleahopper are light green and peppered with small brown spots.

The whitemarked fleahopper is about the same size as a minute pirate bug and has a similar color pattern, but it is less flattened and has longer antennae and enlarged hind legs.

The western plant bug is black with numerous short, silvery bristles.

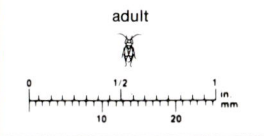

COTTON FLEAHOPPER

adult

0   1/2   1 in
                  mm
10   20

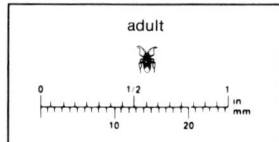

WHITEMARKED FLEAHOPPER/
WESTERN PLANT BUG

adult

0   1/2   1 in
                  mm
10   20

$$\frac{\text{total squares in four 40-inch}}{100} \text{ x } 4 = \text{threshold}$$

The whitemarked fleahopper, *Spanagonicus albofasciatus*, and the western plant bug, *Rhinacloa forticornis*, are small, dark colored plant bugs widespread in the West but most commonly found on cotton in Arizona. The whitemarked fleahopper is black with a white band near the base of the wings; the base of the male antenna is enlarged, and hind legs of both sexes are enlarged. The western plant bug is black with numerous tiny, silvery bristles on the back and wings. Whitemarked fleahoppers occasionally move into cotton from weeds. They feed on cotton terminals to some extent, but they also feed on immature whiteflies and rarely cause measurable loss. Little is known about the habits of the western plant bug, but it is probably more beneficial than harmful; a closely related species is a major predator of *Heliothis* eggs in Peru.

Other bugs occasionally found on cotton include the superb plant bug, *Adelphocoris superbus*; bordered plant bugs, *Largus* spp.; milkweed bugs, *Oncopeltus fasciatus* and *Lygaeus kalmii*; and leaffooted bugs, *Leptoglossus* spp. These bugs are much larger than lygus bugs and/or have markings that clearly distinguish them. They are usually found only in small numbers.

Stink Bugs

Several species of stink bugs occur on cotton in the western U.S.; all have similar life histories and cause the same kind of damage. Stink bugs pierce young bolls to feed on developing seeds. The seeds collapse and the lint is matted and stained. Stink bugs may also introduce bacteria and fungi that cause boll rots. Small bolls injured by stink bugs may drop, and older bolls develop hardened, dry locks which cannot be harvested.

Most stink bug damage to cotton is done by adults that fly in from weeds or from other crops. Large numbers may build up in seed alfalfa, sorghum, or grains, or on weeds such as Russian thistle, and when these hosts mature, the bugs may move to cotton, usually in June or later. Adult stink bugs, easily recognized by their shieldlike shape, are generally green or brown, often with yellow or orange legs and antennae. They range in length from about ⅜ to ½ inch (8 to 15 mm). Among the most common species in the San Joaquin Valley are the Say stink bug, *Pitedia (Chlorochroa) sayi*, and the consperse stink bug, *Euschistus conspersus*. The western brown stink bug, *E. impictiventris*, is similar to the consperse stink bug and is common in the desert valleys. The conchuela, *P. ligata*, is another large, brown species that occurs is eastern Arizona and New Mexico; the edges of its body are bordered with red. Several small species, such as *Thyanta pallidovirens*, are known as red-shouldered stink bugs; they

are green with a red line on the thorax between the bases of the wings.

Although stink bugs usually reproduce on other hosts, the immature stages are found occasionally in cotton. The barrel-shaped eggs are laid in clusters. Nymphs are nearly round and often brightly colored; they remain close together at first but scatter as they grow.

The most important natural enemies of stink bugs are parasitic wasps, *Telenomus* spp., that attack the eggs, and parasitic flies that attack nymphs and adults. Such predators as bigeyed bugs and collops beetles may also attack the eggs and nymphs.

It is seldom worthwhile to monitor for stink bugs routinely, but you should be alert for them, especially along the edge of the field closest to crop or weed hosts. If bugs are numerous, they may appear in the sweep net while you are monitoring for lygus bugs or you may notice the acrid odor they release when disturbed. Also, you may find brown stains caused by stink bug feces on green bolls. These stains, lighter in color than spots caused by lygus bugs, are not sunken. Cut some of the stained bolls to look for the shriveled seeds and stained lint typical of stink bug injury.

Don't confuse stink bug injury with that caused by pink bollworm. Stink bugs produce warts on the inner surface of the boll wall, but the warts are dull white and irregular, while pink bollworm warts are translucent and round. Stink bugs often leave a cluster of warts in the same lock. Staining of lint due to pink bollworm is usually close to the seeds; when the staining is obvious, either a larva or an exit hole is present. In bolls damaged by stink bugs, staining may extend throughout the boll.

The control action guideline suggested for California is to treat when you can find a total of 20 to 25 adult bugs by searching six or seven randomly chosen plants; look for the bugs under the bracts of green bolls. This guideline applies only until early September; although the bugs may be present later, they will not be feeding on bolls. The guideline for New Mexico is to treat when there are two or more stink bugs per 100 sweeps *and* 10% or more of the bolls have symptoms of stink bug injury. The Arizona guideline is to treat when you find 10 to 12 bugs per 100 sweeps.

A strip application of insecticide is often adequate for control of stink bugs, since damaging infestations are usually limited to the edge of the field closest to the previous host plants.

STINK BUGS

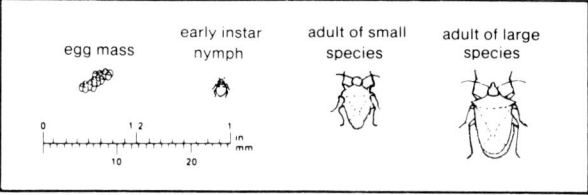

| egg mass | early instar nymph | adult of small species | adult of large species |

The barrel-shaped eggs of stink bugs are laid in groups. Colors vary according to species.

The shieldlike shape of the consperse stink bug is typical of most adult stink bugs.

Stink bug nymphs, such as this Say stink bug, are nearly round.

Stink bugs may cause extensive staining of lint.

Stink bug feeding produces warts on the inner surface of the boll wall. Unlike warts caused by pink bollworms, these are irregular in shape and often clustered.

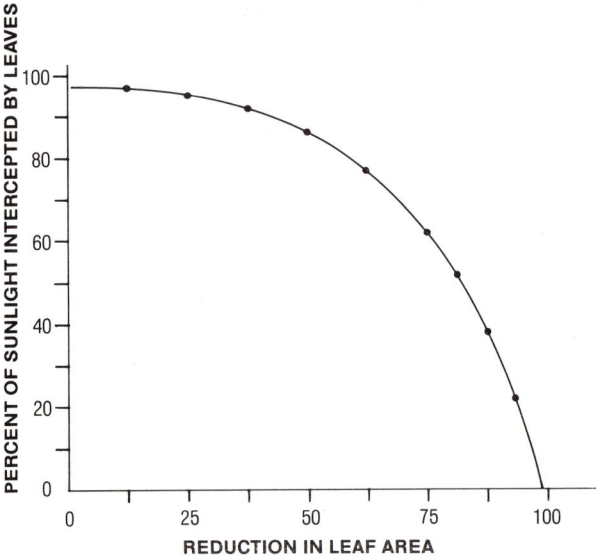

Figure 36. Once cotton has closed over the rows, each plant has a total leaf area three to four times greater than the soil surface area covered by the canopy. For this reason, it takes a large reduction in leaf area to produce a significant drop in the plant's ability to intercept sunlight. For example, a loss of 50% of the leaf area reduces light interception only about 12%.

Female strawberry mites have a large, dark blotch on each side of the body. The egg (lower left) is like a translucent droplet. Twospotted and Pacific mites closely resemble strawberry mites.

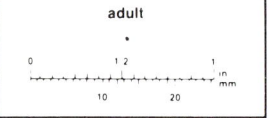

Foliage Pests

Injury to foliage reduces yield only if it is extensive enough to reduce the plant's ability to conduct photosynthesis and transfer energy to fruit. The foliage pests most damaging in most areas are spider mites; their populations multiply so rapidly under favorable conditions that they can injure a large proportion of the foliage quickly. Several species of caterpillars, including beet armyworm, loopers, and the saltmarsh caterpillar, are common, but their injury usually occurs too late in the season to affect yield. After the canopy has closed, plants normally have more foliage than is needed to absorb all available sunlight, so loss of leaf surface does not reduce photosynthesis significantly unless it is extensive (Figure 36). Whiteflies and aphids cause little direct loss of yield by feeding on foliage, but they produce a sticky, sugary honeydew that contaminates lint.

Spider Mites
Tetranychus spp.

Spider mites found on western cotton include the strawberry spider mite, *Tetranychus turkestani*; the twospotted spider mite, *T. urticae*; the Pacific spider mite, *T. pacificus*; the carmine spider mite, *T. cinnabarinus*; and the desert spider mite, *T. desertorum*. In the San Joaquin Valley, the strawberry, twospotted, and Pacific mites are common. In the desert valleys, the predominant species are the strawberry mite and the carmine mite. All species except the Pacific mite are found in eastern Arizona and New Mexico.

Leaves turn yellow or red when injured by spider mites and they may drop when injury is severe. The damage reduces their ability to supply energy to fruit, and may lower lint yield and quality. The strawberry mite, the most destructive species, usually appears earlier in the season than the others and is more likely to cause defoliation and yield loss.

Infestations often begin with mites carried by wind from infested crops such as alfalfa. Outbreaks often follow application of insecticides to cotton. In the absence of insecticides, mites are often kept under control by predators, especially thrips and bigeyed bugs. A sequential sampling plan for the San Joaquin Valley uses the presence or absence of mites on specific leaves to determine the need for treatment.

Description

Spider mites appear to the naked eye only as tiny, moving dots, although you can see them easily with a 10x hand lens. Adult females, the largest forms, are about 1/100 inch (0.25 mm) long. Spider mites live in colonies, mostly on the lower surfaces of leaves; a single colony may contain thousands. The name 'spider mite' comes from the silk webbing some species produce on infested leaves.

Adults have eight legs and an oval body, usually with two red eyespots near the head. The legs and body have numerous long bristles. Eggs are spherical and translucent, like tiny droplets. Immatures resemble adults and feed on leaves in the same way. The first instar, or larva, has six legs; later instars, called nymphs, have eight.

Immatures and males of all spider mites on cotton are superficially identical, but adult females of some species have characteristics recognizable in the field. Females of the strawberry, twospotted, and Pacific mites are green or straw yellow with dark blotches at the sides of the body. Adult females of the carmine mite are bright red. Desert mite females are darker red and are larger than any of the other species. These color patterns are variable, and positive identification of spider mites requires examining the male sexual organ (aedeagus) under a microscope; this is best done by an experienced person with access to reliably identified specimens for comparison.

You do not need to distinguish all the species of spider mites to manage them in cotton, but learn to recognize early infestations of strawberry mites. Strawberry mites on cotyledons and early true leaves form compact colonies on the lower surfaces. Red blotches appear on the upper surfaces opposite the position of the colony. The infested part of the leaf usually puckers upward, and the entire leaf may be distorted. Infested leaves later turn red, then brown, and finally drop. Plants infested early may lose most of their lower leaves by first bloom. Later, the mites spread more rapidly, so infested leaves become generally reddened, lack discrete blotches, and are not distorted. Strawberry mites produce little webbing.

Other mite species may also be present early, but they are usually much less abundant than strawberry mites until later, when their injury appears mostly on mature leaves.

Colonies of twospotted and carmine mites start in a leaf fold or at the base of a leaf blade near the petiole. They gradually spread to the edges of the leaf, producing a diffuse pattern of red and yellow flecks. Leaves drop only when the infestation is unusually severe. Twospotted and carmine mites often produce noticeable webbing on heavily infested leaves.

Pacific and desert mites spread out from the base of the leaf along the main veins, so the yellow pattern of injury is first concentrated there. Once the mites have spread over the entire leaf, injury is similar to that caused by twospotted mites. However, Pacific mites produce more webbing than do other species and they are the only mites commonly found on both surfaces of infested leaves. Desert mites produce little webbing.

Early in the season, strawberry mite injury appears as red blotches on cotyledons or early true leaves.

The mite colony is on the underside of the leaf.

Twospotted mites produce a diffuse pattern of yellow spots over much of the leaf surface.

Strawberry mites can cause extensive defoliation. Note the dead leaves at the base of the plants.

Larva of the western flower thrips feeding on spider mite eggs.

Pacific mite injury often spreads first along main leaf veins, although it may later cover most of the leaf.

From a distance, mite infestations often appear as reddish spots in the field.

Although the pattern of injury varies among species, differences are most obvious early in the season. Late-season infestations often involve more than one mite species and may expand so rapidly that typical patterns are not apparent. Several other conditions produce red or yellow discoloration superficially similar to mite injury. These include air pollution injury, potassium deficiency, repeated exposure to methomyl or to certain herbicides, and injury caused by leafhoppers or bean thrips.

Seasonal Development

Spider mites are present all year on perennial hosts such as alfalfa, and they build up on annual crops and weeds as these become available at various times of year. A generation can take as little as 5 days in hot weather; there are about ten generations each season on cotton. Populations increase fast under favorable conditions.

Because of the wide host range and continuous development, there is always a reservoir of mites that can infest cotton. Infestations usually start with adult females carried by wind from alfalfa, weeds, or other hosts. They often build up first along the edge of the field closest to such sources, then spread in a downwind direction. Infestations may also spread along rows as mites are carried on equipment or clothing that contacts infested plants.

Factors that regulate mite populations are temperature, the condition of host plants, and the activity of predators, especially flower thrips, minute pirate bugs, and bigeyed bugs. Populations increase slowly early in the season while the temperature is low. By mid-June, when nights as well as days turn warm, populations increase extremely fast unless predators are abundant. If there are few predators, there is little chance for them to control the mites, as most predators require a month for a generation, while mites need less than a week. Mite populations usually begin to decline in August as plants channel their energy into maturing fruit rather than into producing new leaves. Only occasionally do populations reach damaging levels after mid-August.

The strawberry mite differs from other spider mites in that infestations generally start earlier in the season, sometimes on seedlings. Strawberry mites occasionally build up even in fields that are not treated with insecticide and that are not close to alfalfa or other infested crops.

Damage

All spider mites cause essentially the same injury; leaves or parts of leaves turn yellow or red and may drop with severe injury. Loss of leaf surface reduces the energy available to maturing fruit, so squares and bolls in affected parts of a plant may fail to develop and may eventually drop. Bolls that do mature may be smaller than normal. A heavy, early infestation may cause plants to lose most of the fruit on lower branches. If the infestation continues, the entire plant may be affected and may produce little or no fruit. Infestations that begin late in the season mainly affect upper leaves; lower bolls that have already matured are not affected and there is little or no loss of yield.

Management

Managing spider mites requires preserving natural controls as long as possible each season and anticipating outbreaks following insecticide applications. The most important enemies of spider mites early in the season are thrips, particularly the immatures or larvae of the western flower thrips, *Frankliniella occidentalis*. Thrips often destroy entire colonies of mites on young plants by feeding on the eggs. In most situations, their value as mite predators far outweighs their injury to young leaves (page 87).

Major predators later in the season include bigeyed bugs and minute pirate bugs as well as thrips. They often keep spider mites below damaging levels, especially when there is an untreated alfalfa field nearby to serve as a reservoir of predators. Predaceous mites and sixspotted thrips also feed on spider mites, but they usually do not build up early enough in the season to control them.

Insecticides often cause outbreaks of mites by destroying predators. There is generally no significant natural control of mites following treatment for lygus bugs or other insect pests. Some insecticides also contribute to mite outbreaks by stimulating egg production. For example, mites exposed to methyl parathion or dimethoate in the laboratory reproduce many times faster than untreated populations. Carbaryl and monocrotophos (Azodrin) can apparently have similar effects. In addition, carbaryl, some organophosphates, and some pyrethroids apparently favor mites by changing the physiology of plants in a way that makes them more suitable as hosts. Insecticides may also favor mites by increasing the level of nitrogen in leaves. Insecticides applied during hot weather usually appear to have the greatest effect on mites, as they can stimulate dramatic outbreaks within a few days. However, insecticides applied earlier can have just as great an effect, although the outbreak may be delayed until temperatures increase.

Monitoring for Mites in the San Joaquin Valley. The critical period for monitoring spider mites in the San Joaquin Valley is from crop emergence until late July, or until about one week after peak square. The sampling plan used in monitoring is based on the gradual upward movement of mites on infested plants. Early in the season, the plants

SEQUENTIAL SAMPLING: SPIDER MITES
threshold = 50% infested leaves
error rate = 20%

leaf number	don't treat	leaves infested	treat	predators (check if present)
1	—	____	—	____
2	—	____	—	____
3	—	____	—	____
4	—	____	—	____
5	—	____	—	____
6	—	____	—	____
7	—	____	—	____
8	—	____	—	____
9	—	____	—	____
10	3	____	7	____
11	4	____	7	____
12	4	____	8	____
13	4	____	9	____
14	5	____	9	____
15	5	____	10	____
16	6	____	10	____
17	6	____	11	____
18	7	____	11	____
19	7	____	12	____
20	8	____	12	____
21	8	____	13	____
22	9	____	13	____
23	9	____	14	____
24	9	____	15	____
25	10	____	15	____
26	10	____	16	____
27	11	____	16	____
28	11	____	17	____
29	12	____	17	____
30	12	____	18	____

DIRECTIONS

1. Choose sample plants *at random*. Don't single out tall plants or plants with visible damage.
2. Each sample is one leaf from the *main stem*.
3. Pick one leaf from each sample plant. Walk at least 20 paces between plants.
4. When plants have less than 9 mainstem leaves, pick the lowest one. When plants have 9 or more mainstem leaves, pick the 8th leaf from the top, counting the newest partly unfurled leaf as no. 1.
5. Use a hand lens to look for mites on the entire lower surface of each sample leaf. Keep a running tally of infested leaves in the center of the card. Don't count a leaf as infested unless you see mites, even if the leaf has symptoms of earlier mite injury.
6. If you find predators, place a check in the appropriate column. Predators are immature thrips, minute pirate bugs, and bigeyed bugs.
7. Treat if the tally of infested leaves reaches the number in the "treat" column. Treatment is not needed if the tally matches the number in the "don't treat" column. You must check at least 10 leaves to make a decision. If the tally remains between the two columns after 30 leaves, sample again in 2 days.

Figure 37. Sequential sampling card and directions for sampling spider mites in the San Joaquin Valley.

are usually growing faster than the mites move, so the uppermost leaves are not infested. Later, as growth slows, mites may reach the top leaves. These movements follow a definite pattern, so it is possible to predict which leaves are most likely to be infested at a given time in the season.

Starting soon after crop emergence, monitor for spider mites once a week. Use one sample card (Figure 37) per field in fields that are fairly uniform. In fields where plant growth varies due to differences in soil type, drainage, or other factors, use separate cards and make treatment decisions separately for each area. Also, use a separate card for "hot spots" close to upwind sources of spider mites or where mites tend to recur each year. Walk at least 50 paces into the field before sampling. Don't sample at the edge of the field unless you plan to treat the edge separately.

Each sample consists of one mainstem leaf from a randomly chosen plant. The exact leaf to pick depends on the size of the plant. When plants have less than nine mainstem leaves, pick the lower one. When plants have nine or more mainstem leaves, pick the eighth to tenth one down from the newest partly unfurled leaf (Figure 5, page 12).

Use a 10x to 20x hand lens to check for mites on the

entire lower surface of each sample leaf. Look in folds where mites may be hidden. If you find mites, tally the leaf as infested. If you do not see any mites, do not tally the leaf as infested even if it is discolored or has webbing or other signs of previous mite injury; mite colonies are often destroyed by predators.

While looking for mites, watch for thrips, minute pirate bugs, and bigeyed bugs. You need not tally the predators, but their presence or absence can help in making treatment decisions when the mite population is close to the threshold.

The treatment threshold for spider mites, 50% infested leaves, is relatively conservative. Some researchers believe that economic injury does not begin until more than 80% of the leaves are infested. However, the 50% level provides a margin for safety, allowing more time to react to sudden increases in mite populations.

When you sample for mites, figure the percentage of infested leaves by dividing the number infested by the number sampled. Plot the results in chart form (Figure 38). If the percentage of infested leaves is increasing rapidly from week to week, sample more than once a week. For example, if the proportion of infested leaves has reached 40% after increasing by 10% or more during the previous sample period (Figure 38, A), then sample again in 3 or 4 days rather than waiting another week. Don't treat, however, until the percentage of infested leaves actually reaches the threshold.

In most fields, monitoring for spider mites is not needed after July. However, if the percentage of infested leaves remains high in late July, especially if it is increasing, continue monitoring until the population declines or stabilizes at a lower level. Regardless of the population level, monitoring and treatment are not needed after most bolls have matured.

Monitoring for Mites in Other Areas. There are no treatment thresholds or specific monitoring procedures in southern California, Arizona, or New Mexico. Be alert for symptoms of mite injury, especially after applying insecticides in summer, but don't treat for mites without examining leaves to make sure mites are actually present.

Control. The acaricides (miticides) most often recommended for use on western cotton are sulfur dust, dicofol (Kelthane), and propargite (Comite). Unlike most other acaricides registered for use on cotton, these are relatively selective; effective for spider mites when properly applied, they have little impact on natural enemies. Sulfur dust is effective for strawberry and desert mites if it is applied when the temperature is 95° F (35° C) or above. It suppresses populations of other spider mites but does not provide adequate control. Because it is explosive and is

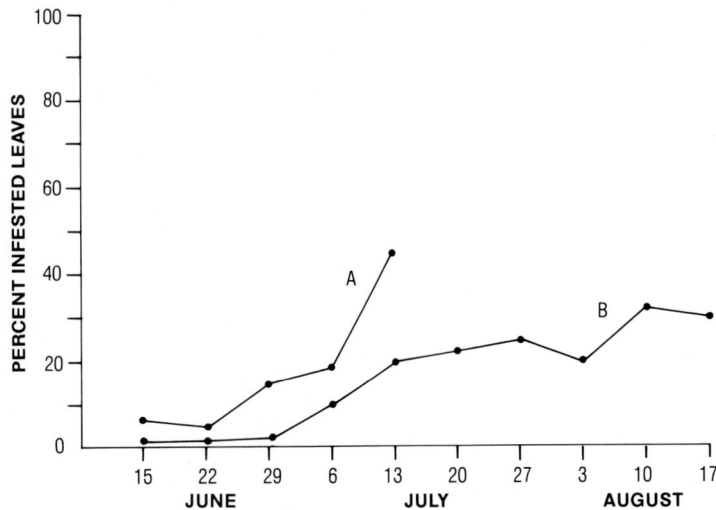

Figure 38. As part of monitoring for spider mites, chart the percentage of leaves found infested in each week's sampling. If the percentage reaches 40% after a jump of 10% or more (A), sample again in 2 or 3 days rather than waiting another week. Continue weekly sampling as long as there are no sudden increases (B).

generally considered too hazardous for aerial application, sulfur dust is usually applied with ground equipment.

Dicofol and propargite can be used in both ground and aerial applications, although they are most effective when applied with tractor-mounted equipment. Arrange the nozzles for best coverage of the lower sides of leaves, where most spider mite colonies are located. For best results by air, the spray plane must pass as close to the canopy as possible. Propargite has generally provided better control than dicofol in aerial applications. A single application is usually enough for the season, except where early treatments for strawberry mites are required; then, a second treatment may be needed later, when mite populations have had time to rebuild. In some cases, treating a limited area along the edge of the field is all that is needed to contain an infestation of spider mites.

Some twospotted and Pacific mite populations in the San Joaquin Valley have shown resistance to dicofol. However, the mites are much more resistant to residues on leaves than to direct contact with a spray. Even where resistant populations are present, it is still possible to achieve adequate control with dicofol by using equipment and gallonage that provide thorough coverage of the lower sides of leaves. So far, significant resistance has been found only in scattered locations, but the possibility of greater and more widespread resistance makes it essential to avoid unnecessary treatments. Researchers have developed a chemical test that allows rapid identification of resistant spider mites and are working on a means of using the test in the field on a commercial basis.

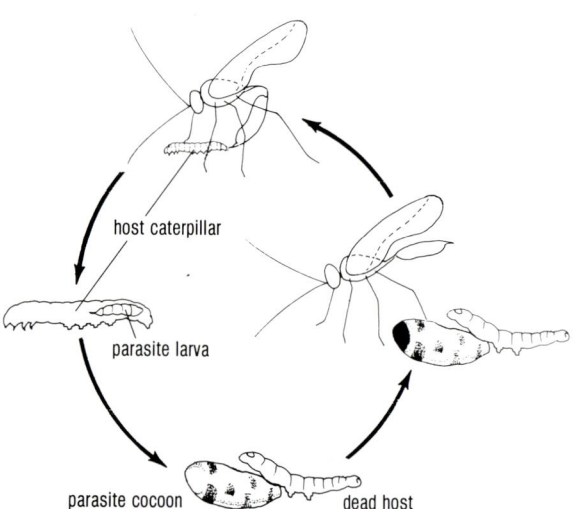

Figure 39. A female *Hyposoter* wasp injects an egg into an early instar caterpillar such as a beet armyworm. The egg hatches into a larva that feeds on body fluids and tissues of the host. The host dies in the third or fourth instar, and the parasite larva emerges from the host's body, spins a cocoon, and pupates. The adult wasp emerges from the cocoon to mate and seek new hosts. Each female may destroy up to 100 host caterpillars. The life cycle takes as little as 15 days, depending on temperature.

An adult *Hyposoter* wasp.

HYPOSOTER

adult

0   1 2   1 in
   10   20 mm

Beet Armyworm
Spodoptera exigua

Fall Armyworm
S. frugiperda

Beet armyworms occur on cotton throughout the Western Region. The fall armyworm, occasionally found on cotton late in the season in the desert valleys, is similar to the beet armyworm in its habits and injury to cotton. The two species can be distinguished by features shown in the Key to Caterpillars, page 55.

Beet armyworm eggs are laid on leaves in clusters covered with hairlike scales. The larvae are dull green with wavy lines down the back and a broader pale stripe along each side. They usually have a dark spot on each side of the body above the second true leg. Young larvae feed together near the egg cluster and gradually disperse as they grow. They skeletonize leaves and bracts, often spinning a loose webbing over the feeding site.

Older larvae chew irregular pieces from leaves and may also feed on squares, flowers, and small bolls. Injury to squares and bolls differs from that caused by *Heliothis* in that it is usually associated with damage to surrounding bracts and foliage. Also, holes are usually in the outer half of the fruit, rather than at the base.

Beet armyworms pupate in a cell on or just below the soil surface. The adult is a mottled gray and brown moth with a wingspan of about 1¼ inch (3 cm). There are three to five generations a year. The pupa is the overwintering stage, but all stages may be present all year in warm areas.

In addition to cotton, beet armyworms feed on alfalfa, vegetables, sugarbeets, beans, and many other crops. Pigweeds, *Amaranthus* spp., and nettleleaf goosefoot, *Chenopodium murale*, are also favored hosts. The parasitic wasp, *Hyposoter exiguae* (Figure 39), is among the most important natural enemies. Populations are often decimated by nuclear polyhedrosis virus.

Damaging populations are most common where insecticides applied for other pests destroy natural enemies. In occasional years, there may be widespread outbreaks when favorable weather allows exceptionally large populations to build up on early-season hosts.

Damage to small bolls can reduce yield if it occurs late enough that plants cannot compensate for it. Feeding on terminals causes excessive branching and may also delay fruiting. In the San Joaquin Valley, plants make up for most beet armyworm damage that occurs before mid-July. In short-season areas such as New Mexico, a delay in fruiting or a loss of bolls at any time may reduce yield. Beet armyworm injury to leaves is important only in rare cases when large numbers of larvae attack small plants. Spotty infestations sometimes occur on seedlings.

There is no treatment threshold, but be alert for large infestations, especially when small bolls are present. To

Beet armyworm eggs are laid in clusters covered with white, hairlike scales from the female moth.

Beet armyworms usually have a black spot on the side of the body above the second true leg.

Young beet armyworms skeletonize leaves and usually spin silk over the feeding site. Early instars feed in a group, but larvae move apart as they mature.

Beet armyworm injury to squares, flowers, and bolls can usually be recognized by the extensive feeding on adjacent bracts and leaves.

BEET ARMYWORM

egg mass   last instar larva   pupa

0   1/2   1
in.
mm
10   20

survey for larvae, use a beating sheet or sweep net. A beating sheet is a piece of cloth or canvas about 3 feet (1m) square. Place it between the rows and shake the foliage of adjacent plants to dislodge larvae; they will fall onto the sheet where you can count them.

Watch for beet armyworms on adjacent crops and on weeds in and around the field. If many larvae are present on weeds while cotton plants are small, it may be worthwhile to use an insecticide to kill them before destroying weeds. Otherwise, they could move to the seedlings and cause a loss of stand. Treatment of a limited area, such as a strip at the edge of the field, is usually successful.

Beet armyworms, loopers, and other caterpillars are often killed by viruses. Dead larvae become black and limp, often hanging from leaves.

Yellowstriped armyworms are dark colored with a broad stripe on each side. They have a black spot just above the spiracle on the side of the first abdominal segment.

Yellowstriped Armyworms
Spodoptera ornithogalli and *S. praefica*

Infestations of yellowstriped armyworms in cotton usually start as outbreaks in alfalfa. When infested alfalfa is harvested or destroyed, larvae may move to cotton or other nearby crops. Only rarely do infestations develop from eggs laid on cotton. When numerous, yellowstriped armyworms can defoliate plants at the edge of a field, and they may damage fruit in the same way as beet armyworms. You can stop migrating larvae by plowing a trench with the steep side toward the cotton, then applying an insecticidal spray or powder to kill the trapped larvae. Baits are also available. Large-diameter irrigation pipe or a strip of stiff aluminum foil set on edge in the soil can also act as a barrier. It is rarely necessary to treat a whole cotton field.

The two species of yellowstriped armyworms are too similar to distinguish in the field. Their life history is similar to that of the beet armyworm.

YELLOWSTRIPED ARMYWORM

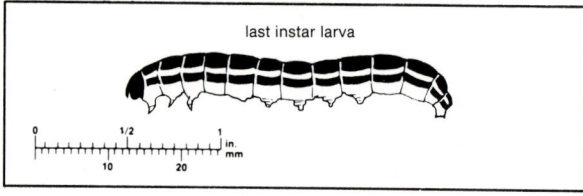

last instar larva

0   1/2   1 in.
           mm
   10   20

CABBAGE LOOPER

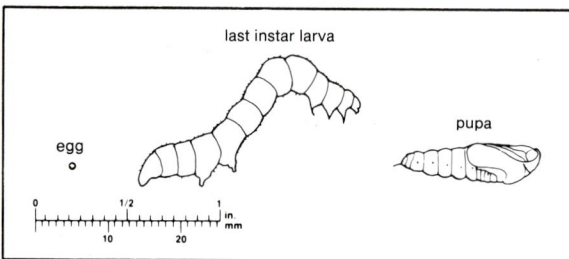

last instar larva

egg

pupa

0   1/2   1 in.
           mm
   10   20

Cabbage Looper
Trichoplusia ni

Alfalfa Looper
Autographa californica

Loopers chew large, ragged holes in leaves but are seldom numerous enough to cause significant damage. They are usually kept under control by natural enemies. The cabbage looper is common everywhere in the Western Region. The alfalfa looper occurs on cotton occasionally in the San Joaquin Valley.

Both looper species are green with several narrow, pale stripes along the back and sides. Easily recognized by their habit of arching into a loop as they crawl, they have only two pairs of prolegs in the middle of the body instead of the four pairs found on most other caterpillars. The brown pupa, similar to that of the bollworm, is formed in a thin silk cocoon on the plant or in debris on the ground. The adult moth has a wingspan of about 1½ inches (3 cm) and has a distinctive silvery spot on the front wing. Looper eggs are similar to those of bollworms, but are more flattened and have finer ridges (Figure 22, page 41). They are laid singly, mostly on the undersides of mature leaves rather than on young terminal leaves where most bollworm eggs are laid.

Loopers have three or more generations a year in most areas. They feed on many weeds and on such crops as alfalfa, beans, and vegetables as well as on cotton. In the desert valleys, development continues all year if host plants are available; where winters are colder, loopers overwinter as pupae. Loopers are most common in July and August, although spotty infestations sometimes occur on seedlings.

Moderate populations may be more beneficial than harmful, as they support populations of natural enemies that also attack bollworms, budworms, and beet armyworms. The eggs and small larvae are attacked by bigeyed bugs, pirate bugs, and other predators. *Trichogramma* parasites kill the eggs, and several other parasites, especially *Hyposoter exiguae* and *Copidosoma truncatellum*, attack the larvae. Loopers are also subject to a nuclear polyhedrosis virus that can reduce populations rapidly.

Ragging of leaves by loopers is not important unless it is very extensive and occurs when it will affect the supply of energy to fruit. During the most vulnerable period, from first square to first open boll, plants can lose 20 to 25% of their leaf area without a reduction in yield. Before and after this period, plants can tolerate a loss of about half of the leaf area. In most cases, control is needed only where insecticides applied for other pests have destroyed natural enemies; in the San Joaquin Valley, cabbage loopers often increase following treatment for lygus bugs. Occasionally, however, widespread outbreaks occur, apparently due to unusually favorable weather.

When loopers are numerous, ragging of leaves is usually obvious, and you will probably find the larvae while sweeping for lygus bugs or monitoring for other pests. You can also find loopers with a beating sheet, as described for beet armyworm. However, there are no specific treatment thresholds for them. If you find large numbers, check for signs of virus disease before applying an insecticide. Insecticides used for control, including *Bacillus thuringiensis*, are effective mainly against young larvae. Spot treatments are usually adequate for infestations on seedlings.

Saltmarsh Caterpillar
Estigmene acrea

Saltmarsh caterpillars are occasional cotton pests everywhere in the Western Region. They cause the same injury to leaves as beet armyworms, skeletonizing leaves in early instars, then ragging the leaves later. Extensive defoliation can reduce yield if it occurs before bolls mature, but heavy infestations seldom occur until late in the season, when their feeding may actually benefit the crop by opening the canopy and reducing the chance of boll rot. Migrating caterpillars that leave cotton fields

Cabbage looper eggs are usually laid on mature leaves.

Cabbage loopers arch their backs as they crawl.

Loopers parasitized by the wasp, *Copidosoma truncatellum*, curl into an 'S' shape after spinning the cocoon, and fail to pupate. Numerous small wasps emerge from each larva.

Eggs of the saltmarsh caterpillar are laid in clusters of a single layer, with no covering of scales. Individual eggs are nearly spherical, unlike drum-shaped stink bug eggs, also laid in clusters.

Early instar saltmarsh caterpillars stay close together, skeletonizing leaves. As they grow, larvae move apart and chew large holes in leaves.

Saltmarsh caterpillars are the only common caterpillars on cotton that have conspicuous tufts of hairlike bristles on each segment.

A parasitic fly approaches to lay an egg on a saltmarsh caterpillar.

often do more damage to nearby vegetables or other crops than they do to cotton. Natural enemies usually control populations in cotton.

Cotton Leafperforator
Bucculatrix thurberiella

The cotton leafperforator, a pest only in Arizona and southern California, can cause severe defoliation. Outbreaks are most common where multiple treatments for pink bollworm destroy natural enemies. Insecticide treatments must be carefully timed to coincide with the brief period when larvae are exposed. The perforator is resistant to many insecticides.

Description and Seasonal Development

Eggs are laid on leaves but are rarely seen because of their small size. Upon hatching, the larva bores into the leaf and tunnels between the leaf surfaces for 3 to 4 days as it passes through the first three instars. The fourth instar emerges from the mine and begins skeletonizing the leaf. After feeding for a day, it constructs a thin silk shelter in which it molts to the fifth instar; this inactive, molting stage is called a "horseshoe" because the larva curls into a "U" shape inside the shelter. The fifth instar is the most damaging, skeletonizing the leaf for 2 to 4 days.

Leafmining instars are flattened and yellow to orange; you can distinguish them from the maggots of leafmining flies by the head capsule and mandibles visible under a microscope on perforator larvae. The fourth and fifth instars are green to gray, with two black spots and several smaller white spots on each segment.

Pupation occurs in a white, ribbed cocoon, usually surrounded by a "fence" of upright silk threads. The adult, a tiny, white moth, emerges in about a week.

Early instar larvae of the cotton leafperforator tunnel between the surfaces of leaves, creating a narrow, winding mine.

Leaves damaged by cotton leafperforators have numerous "windows."

Fourth and fifth instars skeletonize leaves, leaving only the membrane on one side.

The fourth instar spins a loose, silk shelter where it curls into a horseshoe shape while molting to the fifth instar (above). The white cocoon (below) has lengthwise ribs.

The adult cotton leafperforator is a small white moth with a tuft of long scales on the head.

A generation takes only 3 weeks in summer; there may be several overlapping generations per season. Populations are usually kept under control by several species of parasitic wasps that attack the pupae and the leafmining instars. When the parasites are destroyed by insecticides, populations can increase very fast. Damaging populations seldom appear before late July. The perforator feeds on wild cotton, *Gossypium thurberi*, but planted cotton is its main host. Winter survival is greater where cotton is stubbed or abandoned.

Damage

Damaged leaves have numerous small "windows"—holes with a transparent membrane remaining on one side. Heavily infested leaves may be reduced to a network of veins. Most damage occurs in the top third of plants. Severe defoliation may cause bolls to open prematurely, lowering lint quality, and may also cause shedding of squares and small bolls. In rare cases, when the population

COTTON LEAFPERFORATOR

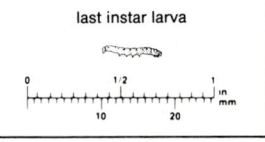

last instar larva

0   1/2   1 in
                          mm
        10   20

is extremely high, larvae may damage bolls directly. Leaves damaged by perforators often fail to drop during chemical defoliation; this increases the trash at harvest and may necessitate extra applications of defoliant.

Management

Infestations of cotton leafperforators usually begin at the edges of a field or in sandy streaks where plants are stressed; watch for damage to upper leaves in such places. There is no specific way to relate the amount of damage to a certain level of yield or quality reduction. A treatment

Eggs of the omnivorous leafroller are flattened and overlap in the cluster, like fish scales.

Omnivorous leafrollers often feed on bracts, fastening them together with silk to form a shelter. The tubercles along the back are white, contrasting with the general body color.

OMNIVOROUS LEAFROLLER

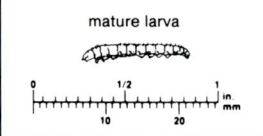

mature larva

guideline suggested in Arizona is to treat when 25 to 50% of the leaves in the top half of the plants have one or more exposed larvae. Count only those larvae on the leaf surface, including horseshoe stage larvae; don't count leafmining instars. The guideline applies only during the part of the season when plants have yet to set a significant part of their boll load. Look for live larvae, not just damage.

Any practice that reduces the use of insecticides lessens the chance of a perforator outbreak. Follow the management guidelines for other pests to avoid unnecessary destruction of natural enemies. Early harvest and plowdown will help reduce overwintering populations.

Timing of treatments is critical, because sprays cannot reach the leafmining instars or horseshoe-stage larvae. Wait until most larvae are in the horseshoe stage; then spray within 2 days to kill the fifth instars when they emerge from their shelters. For heavy infestations, more than one treatment may be required to catch a large enough proportion of the larvae while they are exposed. Strip or spot treatments are adequate for infestations limited to certain parts of a field. Systemic insecticides applied to the soil early in the season do not last long enough to prevent damage in mid- to late summer, when infestations are most common.

Leafrollers, Leaftiers, and Webworms

Several species of small caterpillars found occasionally on western cotton web leaves or bracts together with silk to form a shelter in which they feed. Their injury is sporadic, localized, and seldom economically important. The abundant silk in the shelters distinguishes these species from more damaging caterpillars such as bollworm, pink bollworm, beet armyworm, and cotton leafperforator.

The most common shelter-building caterpillar is the omnivorous leafroller, *Platynota stultana*. Eggs are laid in clusters on leaves or other plant parts. The larvae feed on leaves, bracts, small squares, or on the surface of green bolls; injured bolls may open prematurely. When feeding on bolls, the larvae usually fasten the adjacent bract to the boll with silk. Leafrollers may also tunnel into petioles and small stems, causing leaves and terminals to wilt. When disturbed, the larvae wriggle violently and drop from the plant on a silk thread. Pupation occurs inside the shelters. There are several generations a year.

Leafrollers are most common in mid- to late summer; infestations often spread from seed alfalfa. Control is seldom needed, as natural enemies usually keep populations down. A treatment threshold suggested in Arizona is to treat when 25% of the plants have an active larva.

Apply this guideline only when plants have significant numbers of squares or young bolls that will have time to mature by the end of the season. Control is difficult because the larval shelters protect the pests from sprays.

Other shelter-building caterpillars found on cotton include the beet webworm, *Loxostege sticticalis*; the garden webworm, *Achyra rantalis* (also known as *Loxostege similalis*); the celery leaftier, *Udea rubigalis*; and the false celery leaftier, *Udea profundalis*.

Whitelined Sphinx
Hyles lineata

Larvae of the whitelined sphinx are large caterpillars, up to about 2½ inches (6 cm) long. Because of the distinctive horn on the eighth abdominal segment, they are also known as hornworms. Most common in the low deserts, they feed mostly on native plants, but during occasional outbreaks, large numbers may migrate into cotton fields and defoliate plants. You can stop migrations with a ditch or other barrier, as suggested for yellowstriped armyworms.

Whiteflies

Whiteflies, tiny, sap-sucking insects related to aphids, injure cotton by excreting a sugary honeydew that forms a gummy deposit on lint. Whiteflies are normally minor pests, but there have been severe outbreaks of the sweetpotato whitefly, *Bemisia tabaci*, in the desert valleys, due partly to the destruction of natural enemies by insecticides applied to control tobacco budworm and pink bollworm. The sweetpotato whitefly transmits viruses that cause diseases in several crops, including leaf crumple in cotton.

The bandedwing whitefly, *Trialeurodes abutilonea*, is common in most of the U.S. cotton belt, but occurs only occasionally in the San Joaquin Valley, where the most common species is the greenhouse whitefly, *Trialeurodes vaporariorum*. The sweetpotato whitefly is limited to the desert valleys, where it is often found together with the bandedwing whitefly.

Description and Seasonal Development

The football-shaped eggs are laid on the lower sides of leaves. Leaves where many eggs are laid often have a frosty appearance due to wax deposited by the adults. The nearly microscopic first instars or "crawlers" that hatch from the eggs move to a feeding site on a leaf, where they molt, losing their legs and antennae. Later instars are oval and flattened like small scale insects. The larvae feed by sucking fluids from leaves. They extract much more than they digest, excreting the excess as a clear, sugary liquid—honeydew.

The fourth instar is a pupa that does not feed. In most whitefly species, it is only the pupa that has features that permit the identification of the species in the field. Pupae of the three whiteflies common on cotton differ in the number and size of waxy filaments projecting from the body.

The adult emerges from the pupal skin through a T-shaped slit. Adults are usually yellow with a coating of white, powdery wax on the body and wings. Most species have uniformly dull white wings, but the banded-wing whitefly has two dark bands on each wing. Adults feed on plant sap in the same way as larvae, but produce little honeydew; they fly when disturbed and may be carried long distances by wind.

All whiteflies found on cotton have a wide host range that includes many weeds and crops. They breed all year as long as temperature permits, moving from one host to another as new hosts become available and others dry out or are destroyed by harvest or frost. Whiteflies may be present on cotton at any time, but they usually are most abundant in late summer, when open bolls are susceptible to contamination with honeydew.

The sweetpotato whitefly requires about 15 to 20 days for a generation in summer. Most of this time is spent as a larva. Populations can increase up to ten-fold in 2 weeks, but natural enemies, especially parasitic wasps, normally prevent such an increase from continuing for long. When these enemies are destroyed, there is potential for a tremendous population explosion.

Without magnification, adult whiteflies look like white specks on leaves.

WHITEFLY

Greenhouse whitefly adults lack distinctive markings, but pupae have long, waxy filaments.

Adult sweetpotato whiteflies are indistinguishable from greenhouse whiteflies, but the pupae lack projecting filaments.

The bandedwing whitefly has dark bands on the wings. Filaments on pupae are much shorter than in the greenhouse whitefly.

Insecticides used for pink bollworm and tobacco budworm control, especially the synthetic pyrethroids, are apparently involved in outbreaks of sweetpotato whitefly in desert areas. Data from experimental fields near El Centro show that the sweetpotato whitefly increased sharply when pyrethroids were introduced, even though populations of the bandedwing whitefly did not. Mild winters and the stubbing of cotton may also favor whiteflies by allowing a large population to overwinter.

Damage

Because whiteflies feed on the lower sides of mature leaves, their honeydew falls mostly on the upper sides of leaves and on bolls in the lower parts of plants. Heavily infested plants are often blackened due to a sooty mold that grows on the honeydew. Lint may be reduced in grade if it contains too much mold or honeydew. As long as it remains dry, contaminated lint can be harvested and ginned normally; when moist, it becomes gummy and may clog machinery. In mills, where humidity is high, gummy lint may jam looms. The process for removing honeydew from lint is costly, so mills try to avoid sticky cotton.

The sweetpotato whitefly transmits the virus that causes leaf crumple in cotton (page 109). It is also the vector for viruses that cause disease in lettuce, melons, squash, and other crops; these diseases have caused heavy losses in the Imperial Valley during whitefly outbreaks. Extensive feeding by large populations can slow plant growth, as it drains energy resources, but this type of injury is seldom important in cotton.

Management

The key to preventing whitefly outbreaks is to minimize the destruction of natural enemies. The most important natural enemies of whiteflies are such parasitic wasps as *Eretmocerus haldemani*, which commonly destroys 70% to 90% of the whitefly immatures in the Imperial Valley. The female wasp slips an egg between the whitefly larva and the leaf, and its larva kills the whitefly in the pupal stage. The parasite pupates inside the whitefly pupa and emerges as an adult through a round hole. The round emergence hole is distinct from the slit left by emerging whitefly adults and is the most easily recognized sign that parasites are present. Other parasitic wasps that attack whiteflies include *Encarsia formosa* and *E. meritoria*. Predators such as bigeyed bugs, lacewing larvae, and minute pirate bugs also feed on whiteflies; the white-marked fleahopper is an important predator in western Arizona.

There is no specific way to evaluate infestations, but if whiteflies are present in significant numbers, you will probably notice the adults during normal monitoring. If you see adults, use a hand lens to check for immatures on the undersides of leaves about halfway down the plant; look for the round exit holes in pupae that indicate parasites are present. Heavy whitefly infestations are usually obvious because of the sooty mold on leaves and the large numbers of adults in the tops of plants.

Control with insecticides can be difficult. Whitefly populations consist mostly of eggs and larvae on the lower sides of leaves, where they are hard to reach with sprays, especially those applied by air. Even if a spray reaches them, they are partly protected by a waxy coating. Whiteflies have apparently become resistant to many common insecticides.

Mealybugs
Phenacoccus spp.

Mealybugs, small sucking insects related to aphids, occur occasionally on cotton in the desert areas. They live in colonies on stems and leaves, forming dense, waxy, white masses. Like whiteflies and aphids, mealybugs suck plant sap and produce honeydew. Infestations are usully limited to the edge of a field where they move in from weed hosts. However, more general outbreaks of the Mexican mealybug, *Phenacoccus gossypii*, have occurred in the Imperial Valley, and mealybugs sometimes build up in stub cotton. Control measures are rarely necessary.

Leafhoppers
Empoasca spp.

Several leafhopper species feed on cotton in the western U.S., but the only ones that cause signifcant damage are the southern garden leafhopper, *Empoasca solana*, and the potato leafhopper, *E. fabae*. Adults and nymphs of both species feed by sucking sap from veins on the undersides of mature leaves, mostly in the lower half of the plant. Affected leaves may become distorted and leathery and may develop yellow or red blotches, a condition known as "hopper burn." However, these symptoms may also be due to nutrient imbalances or other factors; the most reliable symptom of leafhopper injury is that the veins are swollen and lumpy.

Other leafhoppers on cotton feed between leaf veins. They may cause a light colored stippling of leaves, but they do not cause swollen veins and their injury does not result in yield loss.

The southern garden leafhopper is most common in the desert valleys, migrating to cotton when fields of its main host, sugarbeets, are dried down for harvest. The potato leafhopper occurs on the east side of the San Joaquin Valley. It migrates to cotton, citrus, and other crops from California buckeye, its spring host. Both species have several generations a year.

Lint contaminated with honeydew becomes sticky and may be discolored by sooty mold.

The parasitic wasp, *Eretmocerus haldemani,* is an important natural enemy of whiteflies. Whitefly pupa killed by the parasite has a round exit hole easily distinguished from the T-shaped slit left by an emerging whitefly adult.

The potato leafhopper feeds on leaf veins, causing them to develop irregular lumps.

LEAFHOPPER

adult

Leaves injured by leafhoppers may be distorted. With severe injury, leaves later turn red.

Leaves injured by bean thrips are covered with small black specks of feces.

Immature bean thrips are marked with orange or red. Adults are black and white.

Natural enemies usually keep leafhoppers from building up large populations in cotton. However, when large numbers migrate to cotton from other hosts, severe injury to leaves may cause plants to shed squares and small bolls. Larger bolls may fail to mature, turning soft and spongy. Such severe injury is rare, however.

Treatment may be needed when extensive symptoms appear, but there is no treatment threshold for leafhoppers. Before applying an insecticide, check for swollen, lumpy veins on a sample of injured leaves to make sure the field symptoms are actually caused by leafhoppers. You can survey for leafhoppers with a sweep net, but it is difficult to distinguish damaging species from others that might be present, so it is not practical to base treatment decisions on results of sweeping.

Bean Thrips
Caliothrips fasciatus

Bean thrips occasionally appear on cotton in summer, usually at the edge of a field where they migrate in from preferred hosts such as sowthistle. Heavy infestations cause mature leaves to turn coppery brown or red and lower leaves may drop. The injury resembles spider mite injury, but affected leaves are covered with tiny black specks which are the feces of the thrips. Adults are similar in appearance to other thrips, but the larvae are marked with orange or red. Don't confuse them with flower thrips (page 87) or other species that are more beneficial than harmful. Spot or strip treatments may be needed for bean thrips in some cases.

Leafminers
Liriomyza spp.

Leafminers, the larvae of tiny flies, feed between the surfaces of leaves, creating a narrow, meandering track that appears at first as a light colored line but later may turn brown. These mines resemble the initial stages of feeding by the cotton leafperforator, but leafminers do not make "windows" in leaves. The yellow larvae, about ⅛ inch (3 mm) long, leave the mine when mature and drop to the ground to pupate. Some leafminers are major pests of vegetables and other crops, but they are usually kept under control in cotton by parasitic wasps that attack the larvae.

Grasshoppers

Various species of grasshoppers feed on cotton leaves, most often along the edges of fields close to pastureland, weedy roadsides, or ditchbanks. They seldom cause economically significant injury, although some fields in the foothills around the San Joaquin Valley have been damaged in occasional years when grasshopper populations reached peak levels. Insecticidal baits available for control are most effective when applied to rangelands or other breeding grounds before the grasshoppers migrate to crops.

Seedling Pests

Many insect pests of seedlings are present throughout the season, although they are most noticeable or most damaging during the seedling stage. Significant injury to cotton seedlings occurs only sporadically in most areas.

Cutworms

Cutworms are caterpillars that hide in the soil or under debris during the day but come out at night to feed. In most areas, they are the most common pests of seedlings, injuring them by chewing through stems near the soil line. Damage is usually limited to certain parts of a field and it may recur each season in the same places. Larger infestations sometimes occur when an alfalfa crop or a stand of weeds is plowed under shortly before planting. To prevent such infestations, keep fields free of weeds and cover crops for at least 3 weeks before planting.

Several cutworm species occur in western cotton fields, including the black cutworm, *Agrotis ipsilon*; the variegated cutworm, *Peridroma saucia*; the army cutworm, *Euxoa auxiliaris*; the granulate cutworm, *Feltia subterranea*; and other *Feltia* species. Most fully grown cutworms are about the same size as a mature bollworm. They usually have dull colors that blend with the soil, and they always appear smooth skinned to the naked eye, without obvious hairs, bristles, or bumps on the skin. Some species appear moist or greasy.

Watch for cutworm injury by walking the field during the seedling stage. Pay special attention to the edges of the field and to any low or weedy spots. If you find seedlings that are cut off, look for cutworms under clods and in the soil nearby. If you walk the field early in the morning or search at night with a flashlight, you may find cutworms on the surface. Various insecticides, including some baits, are available for cutworm control. Use spot or strip treatments, preferably with a ground rig, where injury is concentrated.

Thrips

The thrips most common on young cotton plants are the western flower thrips, *Frankliniella occidentalis*; the tobacco thrips, *F. fusca*; and the onion thrips, *Thrips tabaci*. These are all slender, light colored insects not more than 1/16 inch (2 mm) long. Present all season in most cotton fields, they are usually noticed only in cool spring weather, when their feeding causes the leaves of slow growing seedlings to become wrinkled and distorted. This injury is of little importance to cotton in most areas, and plants quickly outgrow it when the weather warms up. Thrips injury tends to be more severe in eastern New Mexico.

Although these thrips feed on leaves, their larvae also feed on spider mite eggs. They are the most important predators of mites early in the season. In most cases, their value as mite predators is far greater than any injury they cause to seedlings. Insecticides applied to control thrips are usually counter-productive, as they tend to promote outbreaks of mites.

Certain thrips, such as the sixspotted thrips, *Scolothrips sexmaculatus*, are primarily predaceous and do little or no feeding on plant tissue. They feed on mites and insect eggs, but they generally occur too late in the season to affect mite damaging populations. The bean thrips (page 86) is uncommon on seedlings but damages mature leaves.

Thrips lay eggs in plant tissue. Larvae resemble adults except for their smaller size and lack of wings, and they usually feed together with adults. Most species pupate in

Like most cutworms, the variegated cutworm curls up when disturbed.

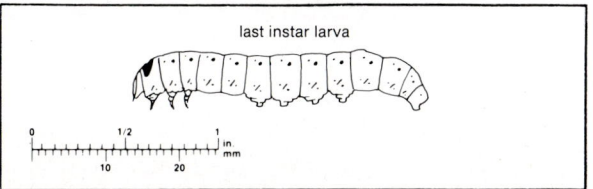
CUTWORM
last instar larva

Young leaves injured by thrips are wrinkled and distorted.

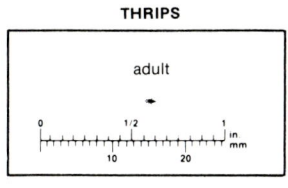
THRIPS
adult

Adult western flower thrips. The larva of this species (photo, page 72) is an important natural enemy of spider mites.

APHIDS

nymph, winged and
wingless adults

parasite

| 0 | 1/2 | 1 | in |
| 10 | 20 | | mm |

Leaves often curl downward when aphids are present on the lower surface.

A typical colony of cotton aphids includes winged and wingless adults as well as immatures.

Aphids killed by parasitic wasps such as *Lysiphlebus testaceipes* turn into swollen, gray or brown "mummies" that remain stuck to leaves. When the adult wasp emerges, it leaves a neat, circular hole in the mummy.

litter or soil beneath the plants. Thrips usually are active all year, moving from one host plant to another.

Aphids

The most common aphid on cotton in the western U.S. is the cotton aphid, *Aphis gossypii*, also called the melon aphid; it may be present any time in the season. Other aphids, such as the cowpea aphid, *A. craccivora*, occur mostly on seedlings. Colonies start with winged females that fly in from other hosts. They produce live nymphs, most of which grow into wingless females. Winged females appear again when the colony begins to decline.

Aphids on seedlings cause crinkling and cupping of leaves. This injury is rarely of economic importance, and plants quickly outgrow it in warm weather. On mature plants, aphids cause the same injury as whiteflies—they produce honeydew which contaminates lint and supports growth of sooty mold. Most infestations are controlled by natural enemies, especially lady beetles and their larvae, syrphid fly larvae, lacewing larvae, and parasitic wasps. If unusually large numbers of aphids build up in certain parts of a field and appear to be retarding growth, apply an insecticide as a spot treatment. Treatment of entire fields, especially preplant application of systemic insecticides, is not recommended. Excessive or poorly scheduled nitrogen applications that stimulate late-season growth favor late infestations of aphids.

Darkling Beetles

Darkling beetles that injure seedlings are oval, brown or black beetles about ¼ inch (6 mm) long. The most common species found in cotton fields are in the genus *Blapstinus*. Like cutworms, they girdle or chew off seedlings near the soil line and are active mostly at night, hiding under clods during hot parts of the day. Darkling beetles sometimes invade cotton fields from weedy areas or from adjacent alfalfa, and large numbers may remain where an alfalfa crop has been disced. Barriers such as those described for yellowstriped armyworms can stop invasions from adjacent fields; otherwise, monitoring and control are the same as for cutworms.

Seedcorn Maggot   *Delia platura*

The seedcorn maggot feeds on seeds and underground parts of young seedlings. The damage has the same effect as preemergence damping-off, as it creates bare areas or areas where the stand is reduced. If you check soon after noticing an emergence problem, you will find hollowed-out seeds or portions of seedlings eaten away. The whitish maggot is similar to a housefly maggot and about ¼ inch (7 mm) long when fully grown. It may be inside damaged seeds or in the soil nearby. Damage is most common in early plantings while the soil is cool, especially in soil that contains a lot of undecayed organic matter. As it seldom affects large parts of a field, control is rarely needed.

Brown Wheat Mite   *Petrobia latens*

The brown wheat mite is a metallic brown mite with long forelegs. It is active only during cool spring weather, when it occasionally migrates into cotton from nearby grain. It does not reproduce on cotton. Injury to seedlings can be severe in rare cases when this mite is present in large numbers, but it is usually limited to the edge of a field closest to grain.

Field Crickets   *Gryllus* spp.

Crickets may feed on cotton plants at any time in the season, but their injury is most noticeable on seedlings. They gouge or girdle the stems of young plants and may also feed on leaves. Feeding occurs at night; crickets hide during the day in soil cracks, ditches, and weeds. Injury is significant only in occasional seasons when exceptionally large numbers build up. Outbreaks are most common in the low deserts. Insecticide baits are available for control. In many cases, treatment is needed only at field edges.

Wireworms

Wireworms are the soil-dwelling larvae of click beetles, family Elateridae. They feed on the roots of many crops and weeds and may bore into the stems of small plants. Injury to cotton is uncommon under western conditions; when it does occur, it appears as spotty stands and poor growth. When wireworms are causing injury, you can find them by searching the soil in the root zone of affected plants. The elongate, cyclindrical, smooth-skinned larvae are yellow or light brown; the small true legs are close together near the head, there are no prolegs, and the tip of the abdomen has a flattened plate. Common species are up to ⅝ inch (15 mm) long. Cultivating, flooding, dry fallowing, and crop rotation can reduce populations of wireworms.

False Chinch Bug   *Nysius raphanus*

False chinch bugs overwinter on weeds, especially mustard family annuals such as shepherdspurse, London rocket, and peppergrass, *Lepidium* spp. When these mature in spring, the bugs may move to other nearby plants, including cotton. In some cases, migrating bugs swarm over seedlings at the edge of a field and injure them severely in a few hours.

The best way to prevent damage is to destroy infested weeds 2 or 3 weeks before cotton emerges; use a sweep net

Darkling beetles are dull bluish black or brown, sometimes with reddish legs. Like *Blapstinus* sp. shown here, most common species have the segments at the tip of the antenna slightly larger than segments at the base.

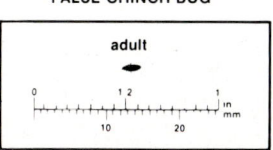

False chinch bugs are about the same size as bigeyed bugs, but they are more slender and the head is more pointed.

to check for the bugs on weed hosts. If you destroy infested weeds after cotton has emerged, it may be worthwhile to treat the weeds with insecticide first to keep the bugs from migrating to the crop afterward. Once an invasion has begun, apply an insecticide at the edge of the field to stop it. There is rarely any need to treat an entire field.

False chinch bugs may be present later in the season, but moderate numbers are of no concern on established plants. Don't confuse these bugs with bigeyed bugs or count them in monitoring for lygus bugs.

Flea beetles produce small pits or holes in cotyledons and leaves.

Notoxus beetles have a hornlike structure on the thorax that extends over the head.

NOTOXUS BEETLE

adult

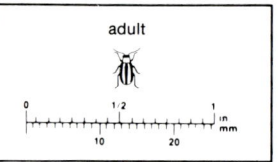

PALESTRIPED FLEA BEETLE

adult

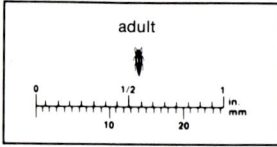

FRUIT BUD BEETLE

adult

The palestriped flea beetle, like other flea beetles, has enlarged hind legs that enable it to jump when disturbed.

Fruit bud beetles feed on pollen but do not injure cotton.

Flea Beetles and Cucumber Beetles

Flea beetles, small beetles with enlarged hind legs that enable them to jump, chew small, round holes or pits in cotyledons and young leaves. They may also be present on older plants, where they feed on leaves and bracts. The most common species in most areas is the palestriped flea beetle, *Systena blanda*. Spotted cucumber beetles, *Diabrotica undecimpunctata*, are slightly larger and are green with a black head and 12 black spots on the wing covers. They do the same kind of feeding on seedlings as flea beetles. Flea beetles and cucumber beetles rarely cause significant injury to cotton.

Miscellaneous Insects

Many insects found in cotton fields have little or no impact on the crop, but it is important to recognize the more common ones so that you will not confuse them with pests.

Notoxus beetles, of which the most common is *Notoxus calcaratus*, are small beetles easily recognized by the hornlike projection of the thorax which extends over the head. Sometimes numerous in cotton, they feed mainly on nectar and do not cause any injury.

The fruit bud beetle, *Conotelus mexicanus*, is a slender, black beetle with short wing covers and knobbed antennae. These beetles are common in the San Joaquin Valley and in the desert valleys, frequently congregating in cotton flowers to feed on pollen. The closely related sap beetles, *Carpophilus* spp., are somewhat smaller and are usually brown with darker markings. They are sometimes found in flowers, but they may also enter holes made in bolls by other insects.

Various species of scarabs or June beetles are also found occasionally in cotton. They are active mostly at night and hide during the day in flowers, where they feed on nectar. *Cyclocephala dimidiata*, a species common in Arizona, is about ½ inch (12 mm) long, shiny reddish brown or black with yellow wing covers.

Nematodes

Nematodes that affect cotton are microscopic, wormlike animals that live in soil and roots. Various species occur in cotton soils, but the only nematodes known to cause economic injury to cotton in the Western Region belong to certain races of the cotton root-knot nematode, *Meloidogyne incognita*. Some nematodes found in cotton soils feed on weeds or rotation crops rather than on cotton, and others feed on fungi, bacteria, or other nematodes.

Injury is most common in sandy soils, where damage to roots may greatly reduce growth and yield. Root-knot nematodes also promote Fusarium wilt. Where soil sampling or experience with nematode injury shows economic loss is likely, soil fumigation may be needed for control. Certain rotation crops can also reduce populations, and field sanitation helps to keep them from spreading.

Description and Biology

Root-knot nematodes are parasites that must feed on roots to reproduce. They feed on a wide variety of crops and weeds. Their feeding causes the host plant to produce a swelling, called a gall or knot, around the feeding site.

Mature females, found only in root galls, are pear shaped and up to 1/16 inch (1.5 mm) long. Under a hand lens, they appear as tiny, white "pearls" inside galls that are cut open. Under normal field conditions, males are rare and mating is not necessary for reproduction. Each female lays up to several hundred eggs in a jellylike mass on or just below the gall surface. Eggs may hatch right away, but they are capable of surviving for long periods when the soil is too dry or cold for growth.

The juveniles or larvae that hatch from the eggs infect new host plants (Figure 40). They can move as far as 2 or 3 feet (60 to 90 cm) through moist soil, attracted by chemicals released from growing roots. Juveniles usually penetrate roots just behind the root tip; the minimum temperature for their growth is about 50° F (10° C), but a soil temperature of about 65° F (18° C) is needed for them to enter roots.

Once a juvenile has entered a root, its salivary secretions cause some root cells to grow abnormally. Several of

Cotton plants injured by root-knot nematodes often have a reduced taproot, as in the plant at right. A normal plant is at left.

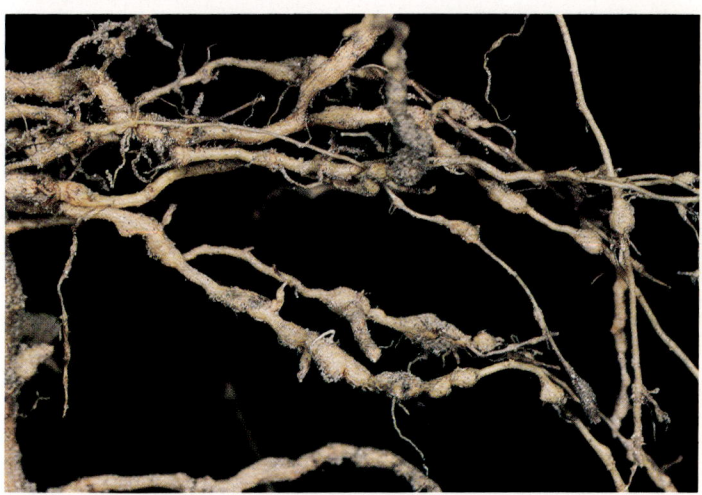

A root system heavily infected with root-knot nematodes has numerous galls. Small lateral roots often branch out from points where galls have formed.

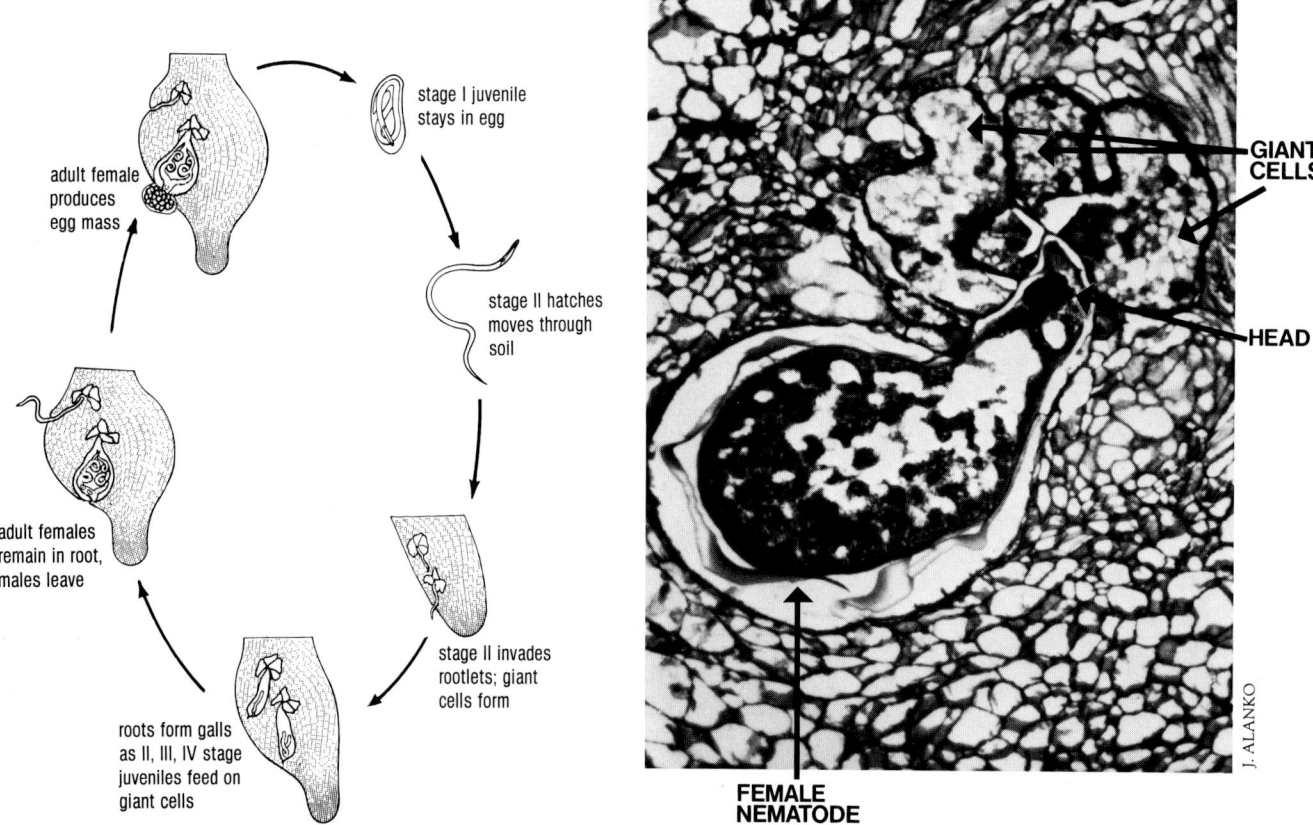

Figure 40. Root-knot nematodes spend most of their life cycle in galls on roots. Second-stage juveniles invade new sites near root tips, and the host plant produces a gall in response to the nematodes' feeding.

Figure 41. A magnified cross section of a root-knot gall shows a female root-knot nematode and several giant cells.

the cells around it enlarge to form "giant cells" that supply it with food (Figure 41). Others multiply to create the gall around the nematode. The nematodes pass through three molts in the roots before they start to lay eggs. As they grow, females swell and lose the ability to move. The life cycle takes about 3 weeks to 1 month when the soil is warm and moisture remains near field capacity.

Root-knot nematode populations increase rapidly during the growing season, but they decline sharply over the winter. Populations in cotton soils typically decline by 80 to 90% between plowdown and planting. Most of the overwintering population consists of eggs and juveniles—a small proportion may survive for a year or more without a host.

Symptoms and Damage

Root galls are the only distinctive symptoms of root-knot nematode injury. Galls on cotton may be up to 1/4 inch (6 mm) in diameter, but they often are much smaller;

they are less obvious than galls on such hosts as sugarbeet or tomato. Infected cotton plants often have a poorly developed taproot and an abnormally large number of shallow lateral roots. Each root gall may have several short roots arising from it. Injury is greatest when plants are infected during the seedling stage, when the taproot is most likely to be affected.

Aboveground symptoms are similar to those of other root diseases and some nutrient deficiencies. Plants may be stunted, especially if infection occurs early in the season, and they may have yellow or reddish leaves. Infected plants may wilt easily, are slow to recover from wilting after irrigation, and respond poorly to fertilizer.

The damaged root system of infected plants limits the uptake of water and nutrients, lowering the plant's capacity for photosynthesis. In effect, root-knot nematodes increase the demand on the plant's energy resources while reducing the supply. Injured plants grow slowly and devote proportionally less energy than normal to producing fruit. As a result, they set relatively few bolls, and these tend to be small. Infected cotton roots are also predisposed to Fusarium wilt (page 103).

Significant losses due to root-knot are usually limited to soils with 50% or more sand—these include sands, loamy sands, and sandy loams. Plants with damaged roots quickly become stressed for water in such soils, and the coarse texture also favors the movement and survival of the nematodes. In finer-textured soils, plants may show little reduction in growth and yield even though root galls are present. Damage may be limited to sandy streaks, where severely injured plants may die and be replaced by weeds by the end of the season. In uniformly sandy soil, damage may occur throughout the field.

Management

To manage root-knot nematodes efficiently, you need to know the cropping history, soil texture, and history of injury in local soils, and you need an estimate of the nematode population level. In sandy soil with a history of root-knot injury to cotton, injury is likely to recur each season as long as cotton or other susceptible crops are planted without soil fumigation. If injury is severe enough and if the affected area is extensive, soil fumigation may be needed or it may be best to plant a resistant crop such as alfalfa before replanting cotton. Field sanitation helps prevent infestations from spreading, and weed control is important in eliminating hosts that support root-knot nematodes.

To make better management decisions, arrange for a nematode analysis of soil samples by a reliable laboratory. By sampling each season, you can develop a record that will show how nematode populations and damage levels change each season following different soil treatments, crops, and cultural practices. Checking for root galls can also provide valuable, though less specific, information.

Checking for Root Galls

Checking for galls on the roots of cotton and other susceptible crops can establish that root-knot nematodes are present and can help to confirm that poor growth or other symptoms are due to root-knot injury. Check at least a few plants each season between midsummer and plowdown, even if the crop appears healthy. Check earlier if you see any abnormal wilting, poor growth, or other symptoms; root galls may appear as early as 1 month after planting.

Draw a simple field map (Figure 42) and dig some plants in each area where plant growth or soil conditions differ; the map will help you plan soil sampling or fumigation should they be needed later. Carefully brush or wash the soil from the roots to look for galls. Be sure to check some plants in the sandiest part of the field, where damage is most likely.

Check roots even if you do not plan to replant cotton the following season, since a root-knot nematode

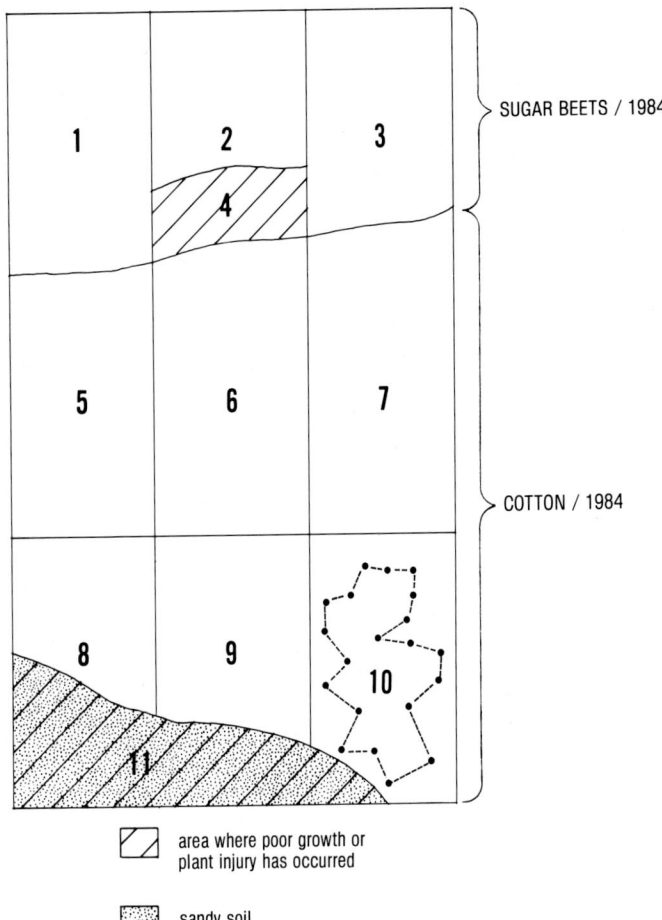

area where poor growth or plant injury has occurred

sandy soil

Figure 42. Before taking soil samples or checking roots for galls, divide the field into areas that reflect any differences in cropping history or soil type. For soil sampling, subdivide these areas into blocks of about 10 acres. If there are areas where there has been poor growth, treat them as separate blocks, even if they are much smaller than other blocks. Assign each block a number and collect soil from a series of points in each one, following a random pattern as shown in block no. 10.

population will persist and may injure cotton in later seasons. In fields where other crops are grown in rotation with cotton, check the roots of those crops also. If the field has been fallowed or planted to a nonhost crop, look for galls on the roots of susceptible weeds such as nightshades and groundcherries. Pigweeds, lambsquarters, and other weeds may also have galls but they are less reliable indicators of root-knot nematode activity.

Because there are several common species, you cannot be certain that galls on other plants are caused by nematodes that would injure cotton. However, if you find any galls in soil where cotton was planted previously, it is reasonable to assume that cotton would be infected if replanted. Unfortunately, the absence of root galls on other plants does not necessarily mean the soil is free of nematodes that could injure cotton.

Soil Sampling

Research on San Joaquin Valley Acala cottons has shown that it is possible to predict the approximate reduction in cotton yield caused by of root-knot nematodes, based on the number of juveniles found in soil samples before planting. As shown in Figure 43, there is a certain preplant population level below which root-knot nematodes have no measurable effect on yield. As the number of nematodes increases, yield gradually declines to about 50% of normal. The relation between yield and preplant population is probably similar in other varieties and other areas, although the numbers may differ. Even though there is no formula for predicting yield loss outside the San Joaquin Valley, a nematode analysis can show whether a population is changing from year to year and can identify parts of a field where root-knot nematodes are concentrated.

Farm advisors can help you find a laboratory equipped for extracting nematodes from soil and identifying them. Contact the lab before you start, since some labs prefer to send their own personnel to collect samples. If you collect the samples yourself, follow the directions below and make sure the lab will be ready to process the samples as soon as they are received. Analysis may take about 2 weeks; sample early enough to allow time for soil treatment if necessary.

You can sample for nematodes at any time of year as long as the soil is in good working condition, but you must take seasonal population trends into account before using lab results to estimate yield loss. Populations of root-knot nematodes reach a peak at about harvesttime, then decline over the winter.

Follow these steps in taking soil samples:

• Draw a map showing any areas in the field that differ in soil texture, cropping history, or crop injury. For example, if the field has a sandy streak, show it as a separate area. Or, if one end of the field was in cotton the previous season while the rest was in corn, treat each area separately. Do the same for places where poor growth has occurred, even it it is limited to a very small area (Figure 42).

• Divide the areas where conditions are uniform into smaller blocks, using a grid pattern. Each block will be sampled separately. If the soil texture and other conditions are uniform throughout the field, apply the grid to the whole field. In general, the smaller the blocks are, the more precise sampling results will be. Most researchers recommend that blocks be no larger than 10 acres.

• Take one sample from each block. A sample is made up of 20 or more sub-samples of soil from different places in the block. Collect the soil with an auger or soil tube that takes a core 1 inch (2.5 cm) or more in diameter. If these tools are not available, use a shovel. Take the soil from the crop's root zone to a depth of at least 18 inches (45 cm); deeper samples are needed in dry, fallowed ground. If growing plants are present, include some roots. Make sure the soil is moist, but avoid places where soil is wet or compacted. If surface soil is dry, discard it and include only moist soil.

• Collect sub-samples from a single block in a bucket or bag. Mix the soil thoroughly, then take out about one quart and put it in a plastic bag or other moisture-proof container.

• Label the sample *in pencil* with the block number. Put the label on the *outside* of the container; moisture may ruin labels placed inside.

• Repeat these steps in each block.

• Keep samples cool; don't leave them in the sun and don't freeze them. The best storage temperature is 50° to 60° F (10° to 15° C). Seal the containers so they won't dry out. A good way to keep samples in good condition is to put them in an ice chest in the field. Remember that the results of a lab analysis can only be as good as the samples.

• Label the whole lot with your name and address, the date, location of the field, crop history, the crop present when you took the samples, the crop you intend to plant, and any notes you have on crop injury.

• Send or deliver samples to the lab as soon as you can, preferable the same day they are collected. Ship them in a cardboard box insulated with newspaper or in a styrofoam ice chest.

Interpreting Sample Results. The methods commonly used to extract root-knot nematodes from soil recover only juveniles. It is not practical to determine their species without rearing them on susceptible plants, a procedure

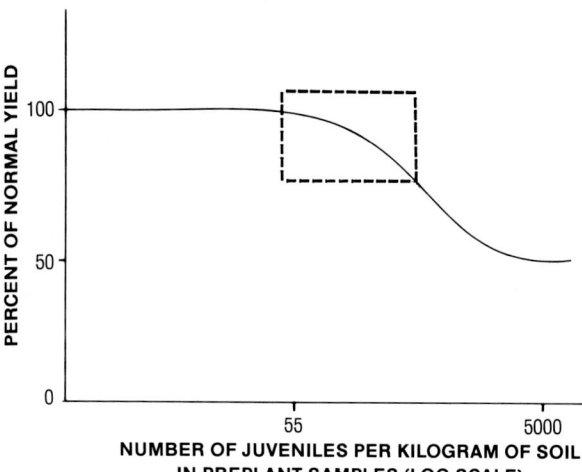

Figure 43. Root-knot nematodes have no effect on yield until the population reaches a certain level. Above that level, increasing numbers of nematodes reduce yield by as much as half. Most treatment decisions are made in the outlined area, which is expanded in Table 10. The equation for this curve is

$$y = 0.479 + (0.521)(0.999^{(P-55)}),$$

where y is the percent of normal yield and P is the number of juveniles per kilogram in preplant soil samples.

that takes too much time for routine use. For this reason, labs generally report the number of root-knot juveniles, *Meloidogyne* spp., found in a certain weight of soil, usually 100 grams or 1 kilogram (kg), without specifying which species are present.

The apparatus most often used for extracting nematodes is the Baermann funnel, which recovers only those juveniles already free in the soil. The addition of a mist chamber promotes the hatching of juveniles from any eggs the samples may contain, so these juveniles also can be extracted; this improves the accuracy of the population estimate. Each lab report should specify the extraction method used and it should also indicate the estimated proportion of the nematodes present that are actually extracted and counted. This figure, called the recovery rate or efficiency, is usually from 10 to 30% for root-knot juveniles.

Control Action Guidelines

For San Joaquin Valley fields, you can use Table 10 to estimate the expected yield loss based on the results of a soil analysis. Table 10 is derived from the same data used to produce Figure 43, but it is limited to the levels of nematode population and yield loss where most economic decisions must be made. To use the table, you must take into account the recovery rate of the extraction procedure; the time of year when the samples were taken; the soil texture in the field sampled; and the cropping history of the field.

If your lab reports an estimated *total* number of juveniles per kilogram of soil, you can compare the results directly to Table 10. However, if the lab reports only the number *extracted* from the samples, then you must divide the number by the recovery rate and multiply by 100 to get the estimated total. For example, if the lab recovered 30 juveniles per kilogram and the recovery rate is 20%, then the total would be (30÷20) x 100, or 150 per kilogram. Labs do not always state the recovery rate in their reports, but they should be able to supply it on request. If the lab uses a unit of weight other than a kilogram, adjust the figures accordingly before comparing them to Table 10.

Table 10 is based on soil samples taken in March in sandy loam soil. You cannot use it to evaluate samples taken in fall, since there is no way to estimate accurately the decline in population over the winter. On finer textured soils such as silt loams or clay loams, cotton can tolerate higher numbers of root-knot nematodes than on sandy loams, so the expected yield reduction for a given population level would be less than that shown in the table. Don't use Table 10 for fields with a history of Fusarium wilt; in the presence of the Fusarium wilt fungus, a relatively small number of root-knot nematodes may cause significant injury.

A yield estimate from Table 10 would not apply to a field where alfalfa has been planted for 2 or more years

before sampling, since many of the root-knot nematodes in the soil would probably be species that do not damage cotton. Other rotation crops have little effect on the interpretation of a soil analysis as long as cotton has been grown in the field in the previous 2 or 3 years. If the field is one where cotton has never been planted, you would need a bioassay (greenhouse test) to determine whether any root-knot nematodes found in soil samples will attack cotton.

Table 10. **Effect of Root-Knot Nematode Population on Yield, As Measured by the Number of Juveniles per Kilogram in Samples of Sandy Loam Soil Taken in March in the San Joaquin Valley.**

Number of Juveniles	Percent of Normal Yield
0 to 55	100
100	98
150	95
200	93
250	91
300	89
350	87
400	85
450	83
500	81
550	80
600	78
650	77
700	75
750	74
800	73
850	71
900	70
950	69
1000	68

The purpose of estimating the expected yield loss is to limit soil treatment to situations where it will improve net return. Base your decision on the cost of soil treatment compared to the value of the yield loss expected from the nematode population level. For example, if a lab finds 200 juveniles per kilogram in sandy loam soil, Table 10 shows that the expected yield will drop to about 93% of normal. If the normal yield is 1000 pounds per acre, then the yield loss is 7% of 1000, or 70 pounds per acre. When cotton is worth 80 cents per pound, the value of the loss is 70 x 0.80, or $56 per acre. If the cost of soil treatment is less than $56, then treatment will increase the net return. Take into account the history of losses in previous crops before making a final decision.

Soil Treatments

To get the best results from soil fumigation, make sure the soil temperature and moisture are right for the material you are using. Before fumigating, check the soil temperature at the depth of application with a thermometer. The best temperature is from about 60° to 77° F (16° to 25° C), although fumigation with dichloropropene (Telone II) can be effective down to about 50° F (10° C). Results of fumigation are generally poor if the soil temperature is above 80° F (27° C), a level not uncommon in desert areas in fall.

The soil should be moist enough that it can be worked easily, but it should be well below field capacity; a level of 40% to 60% of field capacity is generally best. The cooler the soil is, the more important that it not be too wet. Eliminate large clods before treatment if possible, since fumigants will not penetrate them.

All soil fumigants now available are toxic to cotton, so a waiting period of 10 days or more is usually required between application and planting. Follow the directions on the fumigant label and observe the suggested waiting period.

Soil fumigants used in cotton fields are usually applied to beds during listing. Set one or two shanks in front of the listers to inject the fumigant 12 to 18 inches (30 to 45 cm) below the top of the finished bed. If you use two shanks, set them 10 to 12 inches (25 to 30 cm) apart and stagger them so that they won't collect trash. List the soil over the injection site and seal the soil with a ring roller or similar tool.

Although most fumigations are carried out in spring, fall fumigation is also effective as long as soil conditions are suitable at the time of application. Remember, however, that you cannot base a treatment decision on Table 10 if you take soil samples in fall rather than spring.

Nonfumigant nematicides can be applied when soil conditions are unfavorable for fumigation, but they provide much less protection from root-knot injury. Soil-applied systemic pesticides such as aldicarb (Temik) kill some nematodes and temporarily prevent juveniles from entering roots, but they do not provide significant control when applied at the doses normally used for insects. The effect of aldicarb on nematodes is due to direct contact with the material in soil water; its systemic activity has little effect once nematodes have entered roots. Don't rely on an insecticidal application of aldicarb if a nematode analysis or other information shows that a damaging population is present.

Resistant Varieties and Rotation

In managing soils infested with root-knot nematodes, you must distinguish among crop varieties that are susceptible, tolerant, or resistant. In *resistant* varieties, juveniles may enter the roots, but they cannot reproduce. In *tolerant* crops and varieties, nematodes reproduce without causing significant reduction in growth or yield. In *susceptible* crops, they both reproduce and reduce growth or yield.

Most varieties of sugarbeet, tomato, beans, potato, and other susceptible crops are hosts for several species of root-knot nematodes. Cotton is unusual in that it is susceptible only to certain races of one species. Also, cotton can tolerate a higher number of root-knot nematodes without loss than most crops. When planning rotation crops in cotton soils, remember that damaging nematodes could be present and might injure other crops even if cotton shows no symptoms of injury.

No commercial cotton variety grown in the West is resistant to the cotton root-knot nematode; however, resistance is available in some experimental lines, and it may be incorporated into commercial varieties in the future. Some cotton varieties, including Acala SJ2 and SJ5 grown in the San Joaquin Valley, have a limited degree of tolerance.

Alfalfa is the most useful rotation crop for infested soils. Most varieties are resistant to the cotton root-knot nematode, although they may be susceptible to other species. Keeping alfalfa in the field for 2 or 3 years generally reduces the population enough that cotton can be grown without soil treatment for at least one season. Some varieties of blackeyed pea and tomato are also resistant to the cotton root-knot nematode.

Corn, the most widely planted western crop with significant tolerance; may produce a profitable yield where susceptible crops would fail, but it does not prevent root-knot nematodes from increasing. A crop of corn allows approximately the same population increase as a cotton crop. Certain corn varieties are resistant, but they are not generally planted in the West.

Winter crops such as small grains do not allow root-knot populations to increase as long as the crops are planted when the soil temperature is below about 65° F (18° C). In cool soil, few juveniles can enter roots so there is little reproduction. If planted too early, however, a susceptible winter crop could result in a higher spring population than would have been present if the soil were left fallow over the winter.

Summer fallowing can reduce populations if the field is disced occasionally to destroy weeds and to expose nematodes to the air's drying action. Fallowing can be useful in fields where a winter crop is harvested too late to allow planting cotton.

Weed control and soil analysis are essential to any rotation program on infested soils. Even if you plant alfalfa or other resistant crops, populations of cotton root-knot nematode could persist and possibly increase if the field remains weedy; nearly all common agricultural weeds are potential hosts. It is a good idea to take samples for analysis before replanting cotton to watch for any unexpected changes in populations.

Sanitation

Root-knot nematodes can spread in irrigation water and in soil carried on farm equipment. A small infestation in one part of a field can easily spread throughout the field when soil is moved in grading. To avoid spreading infestations, avoid moving infested soil and don't use tail water from infested fields.

Diseases

Each area in the western United States has at least one cotton disease capable of causing significant loss. The major diseases are caused by pathogens—organisms that invade the host plant and disrupt its normal functions. Most of the pathogens are fungi, but bacteria and a virus are involved in some diseases. Nematodes, also considered plant pathogens, are treated separately in this manual. Other disorders are caused by such factors as nutrient deficiencies, environmental stress, or exposure to toxic substances, including improperly applied pesticides.

On a regional basis, the most destructive diseases of cotton are caused by soilborne fungi that invade plants through the roots. These include seedling diseases, Verticillium wilt, Fusarium wilt, and cotton root rot. Seedling diseases occur throughout the region wherever wet soil favors the fungi and especially when low soil temperature slows the growth of cotton seedling. Prevention is based on fungicide seed treatments and on planting cotton when conditions are favorable for germination and seedling growth. Verticillium wilt is a major disease in New Mexico and in the San Joaquin Valley; management is based on crop rotation and tolerant varieties. In the low deserts, the Verticillium fungus is present, but the disease is suppressed by high temperatures. Cotton root rot, on the other hand, is favored by high temperatures and alkaline soils and is most prevalent in the Arizona deserts. No reliable, economical control is available. Fusarium wilt is known only in the San Joaquin Valley, where it is associated with root-knot nematodes.

Aboveground diseases are those in which infection results from fungal spores or bacteria blown or splashed onto plants. These diseases are favored by high humidity, and some require free moisture for development. Boll rots caused by fungi and bacteria occur everywhere in the Western Region. They are most prevalent during hot, humid weather in the low deserts, but they can also cause losses in other areas, especially in fields where rank growth keeps the humidity of the canopy high. One boll rot results in contamination of cotton seed with aflatoxins; this problem is generally limited to the low deserts. Cultural practices that discourage rank growth and that maintain air circulation in the plant canopy tend to reduce boll rots. Control of insects that damage bolls is also important, since damaged bolls are more vulnerable to infection. Other aboveground diseases, including southwestern cotton rust, bacterial blight, and Alternaria leaf spot, seldom cause significant losses; they are most common in New Mexico and southeastern Arizona.

Cotton leaf crumple is the only disease of cotton in the Western Region caused by a virus; it occurs only in the low deserts. The virus is carried by the sweetpotato whitefly. Leaf crumple is most prevalent in stub cotton, but it can spread to planted cotton in seasons when whitefly populations are high.

Monitoring and Diagnosis

Familiarize yourself with the symptoms of common diseases and watch for signs of stress as part of regular field monitoring. Before the season begins, read the sections on specific diseases in this manual and check the illustrations so that you will know what to look for.

Any plant disease, whether caused by a pathogen or not, involves a complex interaction between host plant and environment. The symptoms produced by a disease and the rate at which they develop are influenced by genetic characteristics of the plant, by the stage of growth when infection or stress occurs, by other stresses that may occur at the same time, and by environmental conditions such as temperature and humidity. Some pathogens, such as the Verticillium wilt fungus, include strains or races that can produce different symptoms.

When you see symptoms of stress or disease, take note of any pattern they may have; note whether they appear only on scattered plants or in certain parts of the field, or are generally distributed. Take weather and soil conditions into account; moisture is especially important, since many pathogens are spread by water or require moisture for infection.

When comparing symptoms in the field with illustrations and descriptions, examine as many affected plants as possible. Look for plants showing different stages of disease development to determine how symptoms change as the disease progresses. Don't rely on a single symptom, such as a leaf spot or yellowing, to identify a disease, but check all parts of affected plants, including roots and stems. Different stresses or pathogens may produce similar symptoms if they disrupt the same plant function. For example, soilborne fungi, root-knot nematodes, soil compaction, and improper herbicide applications may all cause stunting because they all interfere with absorption of water and nutrients. A set of several symptoms is usually needed to diagnose a disease.

It is not always possible to identify diseases with certainty in the field. Some pathogens require special laboratory techniques for identification, and nutrient deficiencies and some other conditions may require plant tissue analysis. When laboratory services are needed, accurate field notes can help confirm the results.

Seedling Diseases

Seed decay, damping-off, and root rot of seedlings occur everywhere cotton is grown and are among the most damaging diseases in the Western Region. The pathogens are soilborne fungi. The most common are *Rhizoctonia solani*, *Thielaviopsis basicola*, and *Pythium* spp., especially *P. ultimum*. Losses due to seedling diseases are greatest when germination and seedling growth are slowed by cool, wet conditions. To reduce losses, use high quality seed treated with fungicides and plant when favorable weather is expected. In some cases, it may be worthwhile to apply fungicide to the soil at planting.

Symptoms and Damage

Fungi that infect cotton seedlings can produce a range of symptoms, including decay of seed, lesions or discoloration on the hypocotyl, and destruction of roots. Some fungi produce characteristic symptoms when they are present alone, but seedling diseases often involve more than one species and it is not always possible to tell which ones are present without culturing them in the laboratory.

Preemergence damping-off—the decay of seeds or seedlings before emergence—is most often associated with *Pythium* spp. The disease appears simply as spotty emergence or bare spots. You may be able to find deteriorated seeds or seedlings in the soil if you look soon after normal seedlings have emerged.

Symptoms of postemergence damping-off often include brown or black lesions on the hypocotyl near the soil line. Lesions may be as small as a pinhead or may completely girdle the stem. Relatively large, brown, sunken lesions caused by *Rhizoctonia solani* are often called "soreshins." Seedlings subjected to wind while the soil has a stiff crust may develop an abrasion that resembles this kind of lesion.

Root development in diseased seedlings is usually reduced, and the tap and/or lateral roots may be discolored or rotted. *Thielaviopsis basicola* produces black root rot, a condition in which the cortex of the hypocotyl and tap root turns black while vascular tissue remains white as in normal plants. Infected plants can be pulled up easily and have few or no lateral roots.

Severely injured seedlings may quickly dry out and shrivel. Those girdled by lesions at the soil line may fall over. Other infected plants may survive, however, if growing conditions improve soon enough. Plants infected by *T. basicola* usually produce new cortical tissue when the weather warms up, eventually sloughing off the infected tissue. Symptoms of seedling infection sometimes remain on older plants; black root rot occasionally persists through the season, and plants with injured tap roots may develop a swollen area above the point of injury.

Replanting may be necessary when seedling disease affects a large portion of a field. Losses then include the costs of new seed, herbicide, and labor as well as lost time. There may also be indirect losses, since the crop may then mature later and may not have time to mature the maximum fruit load. More water may be needed, too, both to bring the soil back to field capacity before replanting, and to keep the crop growing later in the season during hot weather.

Even when replanting is not necessary, losses caused by seedling diseases are not limited to spots where plants are killed. Plants that survive infection often have a poor root system. They are prone to water stress and may yield less than normal. The lack of uniformity in a field where some parts are affected by seedling disease can make irrigation and other cultural operations more difficult.

Seasonal Development

Fungi that cause seedling diseases are widespread in western cotton soils. Most have a wide host range that includes many crops and weeds, so they may be abundant even in soil where cotton has not been planted before. In addition to living as parasites in growing plants, they may also live in the soil as decay organisms and they survive for long periods as resistant forms such as spores or sclerotia.

Seedling diseases are most prevalent in wet, cool soils. However, the fungal pathogens vary in their temperature requirements, so infection can also occur in warm soil, especially when soil compaction or other conditions delay emergence. *Rhizoctonia solani* infects seedlings throughout the Western Region. The only seedling disease pathogen that regularly causes losses in Arizona, it is also important in the San Joaquin Valley and New Mexico. *Pythium ultimum* is most active in cool soils and is a major seedling pathogen in the San Joaquin Valley and New Mexico. *P. aphanidermatum* is adapted to warmer soils and causes disease occasionally in the desert valleys. *Thielaviopsis basicola* is active in cool soils and is similar in distribution to *P. ultimum. Fusarium oxysporum* f. sp. *vasinfectum* can be a seedling disease pathogen in sandy San Joaquin Valley soils in the presence of root-knot nematodes.

Management

To reduce seedling diseases, make sure that conditions at planting favor rapid germination and seedling growth. Under favorable conditions, cotton seedlings quickly outgrow the most vulnerable stage and infection is less likely.

Always use the highest quality seed you can afford. If possible, select seed that has shown a high rate of germination in a cold test. If you must use lower quality seed, plant as late as possible to allow the soil to warm up. Regardless of seed quality, never plant if rain or cold weather is expected during the 4 or 5 days following planting. Use an adequate seeding rate so that the loss of a few plants to seedling diseases will not leave skips that must be replanted. Don't plant too deeply, since excessive depth delays emergence and exposes more hypocotyl surface to invasion by fungi.

Soil that is too wet at planting or during germination favors seedling diseases. To avoid excess moisture, allow preirrigated beds to drain adequately before planting, and don't irrigate up during cool weather. Firming wheels on planters operated in wet soil often create a shallow compacted layer that aggravates seedling disease problems. Roots growing through compacted layers may develop constricted, weakened areas vulnerable to infection by fungi and may restrict growth later in the season.

Fungicide seed treatments can usually prevent severe losses due to seedling diseases as long as growing conditions are reasonably good. Always use seed treated with fungicides effective against *Rhizoctonia solani* and *Pythium* spp. In cooler areas, especially in early plantings, it is advisable to include a material effective against *Thielaviopsis basicola*. Supplemental fungicides applied to the soil at planting can provide extra protection. They are often worth the extra cost when cool, wet weather is likely after planting, in fields with a history of severe seedling disease, and in fields that must be replanted due to seedling disease.

Take care in selecting herbicides, insecticides, and other chemicals applied to the soil at planting. Excessive concentrations of these materials can delay emergence, especially in cool weather, and may favor seedling disease. Insecticidal seed treatments may also affect emergence in cool conditions.

Rotations that include grass crops such as small grains, corn, or sorghum can help reduce the inoculum of seedling disease fungi in the soil. The inoculum tends to increase in fields where cotton is grown every year, and it

Seedlings infected by soilborne fungi may collapse rapidly.

The root system of diseased seedlings is reduced or may be entirely destroyed, as in the plant at right. A normal seedling is at left.

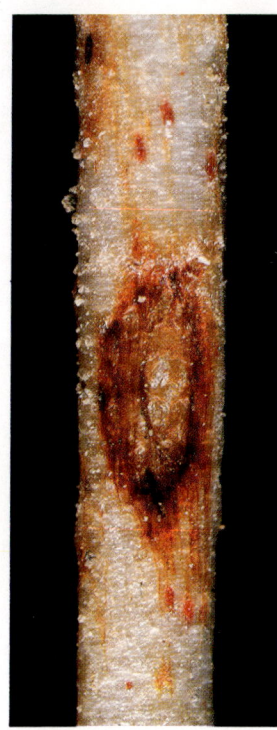

Rhizoctonia solani often causes dark lesions near the soil line.

may also build up in alfalfa, sugarbeets, and black-eyed peas. Where cotton is grown for a long period without rotation, more aggressive strains of seedling disease fungi may increase more than other, less damaging strains.

Verticillium Wilt
Verticillium dahliae

Verticillium wilt occurs in most parts of the world where upland cotton is grown. It is the most damaging disease of cotton in the San Joaquin Valley and New Mexico. In Arizona, it occurs mostly at elevations above 2000 feet; high temperatures usually restrict its development in the low deserts. Losses can be reduced by planting tolerant varieties, by following good irrigation and fertilization practices, and by rotating cotton with grass crops to prevent the inoculum in soil from reaching a damaging level. Pima cottons grown in the Western Region are highly tolerant of Verticillium wilt.

Symptoms and Damage

The first external symptom of Verticillium wilt is an abnormally dark green color of foliage on plants showing poor growth. Leaves may then develop diffuse, yellow patches between the main veins. As these patches expand, tissue at their centers dies and a large part of the leaf may become mottled with yellowed and dead tissue. Severely injured leaves may drop. Although the disease may begin at any time, foliar symptoms seldom appear before squaring.

Verticillium wilt also causes brown streaks in the xylem. Discoloration usually is most obvious in the main stem and branches, but it also appears in petioles and the main veins of leaves, where it is possible to check for it without injuring the plant. Vascular discoloration nearly always appears soon after infection; its presence, however, does not necessarily mean that foliar symptoms will appear later or that yield loss will occur. Tolerant cotton varieties often show extensive vascular discoloration without foliar symptoms or yield reduction.

High temperatures inhibit the expression of leaf symptoms although other aspects of disease development are not affected. For this reason, plants may appear to recover from Verticillium wilt during periods of hot weather as yellowed leaves are hidden by new leaves that do not show any symptoms. These plants do not actually recover, however, but remain less productive than normal even if most of the foliage is still green late in the season. Yellowing of leaves on infected plants may progress rapidly when temperatures drop in late summer.

Highly virulent strains of the Verticillium wilt fungus can quickly defoliate and kill plants of most upland cotton varieties, and highly susceptible varieties may be severely stunted even by milder strains. However, most varieties grown where Verticillium wilt is common are moderately tolerant, so infected plants are seldom killed.

Verticillium wilt restricts growth and reduces yield by interfering with the movement of compounds produced in photosynthesis. In effect, the flow of energy from leaves to squares, bolls, and stems is reduced. Large bolls that are present when foliar symptoms appear often mature, but squares and small bolls often drop, and production of new squares stops. The amount of yield loss depends largely on when foliar symptoms appear; in general, the earlier foliar symptoms appear, the greater the loss.

Symptoms are similar to those caused by the Fusarium wilt/root-knot nematode complex. In the San Joaquin Valley, where both diseases occur, they could be confused. The most reliable characteristic distinguishing the two is that chlorosis in Verticillium wilt is usually patchy and evenly distributed on leaves, while in Fusarium wilt it begins in discrete spots often confined to a single leaf lobe. Also, some lateral roots on plants with Fusarium wilt may be blackened externally, while roots with Verticillium wilt are not discolored even if they have root-knot galls. Both diseases cause vascular discoloration, but the discoloration associated with Verticillium wilt is usually streaked and relatively light in color, while discoloration due to Fusarium wilt is darker and more continuous. Another difference is that the Fusarium/nematode complex is more likely to cause severe stunting or death of infected plants. However, the differences between the two diseases are not always clear in the field. The only reliable way to confirm a diagnosis is to culture and identify the fungus in the laboratory.

Seasonal Development

The Verticillium wilt fungus may be present in soil in several forms. Its mycelium acts as a decay organism in debris from infected plants, and the fungus also produces resistant forms such as microsclerotia that remain viable for years even under adverse conditions. All of these forms are part of the fungus inoculum or reservoir that can infect new host plants.

Infection occurs when a growing root of a susceptible plant approaches a microsclerotium or other form of the fungus in soil. Stimulated by chemicals released from the root, the fungus produces a slender filament (hypha) that penetrates the root and enters the xylem. The filaments grow through the xylem tissues and produce numerous microscopic structures called conidia that are carried upward through the plant by the flow of water in the xylem. In leaves, conidia germinate and cause formation of a gel that blocks circulation of water and nutrients, killing the affected part of the leaf.

The cycle is completed when infected plant tissue returns to the soil. The fungus then produces new microsclerotia and other resting forms that are released into the soil as the infected debris decays. About 80 to 90% of the inoculum is in the top 1 foot (30 cm) of soil.

Management

The concentration or density of inoculum in soil is a major factor in choosing management strategies for Verticillium wilt. Where the density is low, you can generally prevent increases by following a regular rotation with non-susceptible crops, particularly grass family crops such as corn and small grains. Tolerant cotton varieties reduce losses, but they do not prevent inoculum from increasing. Once the inoculum reaches a high level, it may be necessary to rotate out of cotton for several years or employ special techniques such as soil solarization to reduce it.

Monitoring. The number of microsclerotia found in soil samples can serve as an index of the inoculum level. Laboratories that analyze soil for *Verticillium* use either the wet sieving or the Anderson sample technique and report the number of microsclerotia found per gram of soil. The number recovered by the Anderson method averages 2.8 times higher than that recovered by wet sieving, so you must know which technique was used before interpreting an analysis. All numbers in this section refer to wet sieving. The assay techniques were developed for the San Joaquin Valley, but they could also be applied elsewhere.

Since microsclerotia are produced gradually as infected plant debris decays, the number found in soil samples increases after plowdown and peaks in midsummer the following year. To monitor changes in inoculum level from year to year, be sure that soil samples are taken at the same time each year.

Surveys using the wet sieving technique to analyze soil samples taken in July have found that inoculum density in San Joaquin Valley cotton soils varies from near 0 to about 40 microsclerotia per gram; the average level is about 5 to 7 per gram. Where a single cotton variety is planted without rotation, a level of 10 or more microsclerotia per gram usually results in significant yield loss. Similar information is not yet available for other areas.

Soil assays enable you to monitor changes in inoculum density. Where susceptible crops are planted repeatedly, the density usually increases from year to year. The rate of increase is affected by seasonal temperatures, the strain of the fungus present, and other factors. Inoculum density increases gradually in some situations, but in other cases it may increase rapidly to a level that will reduce yields. Samples taken in summer can help you decide whether to plant an alternate crop or a new variety the following season, while samples taken in spring can help in choosing management options in the current season.

When taking soil samples for *Verticillium* assay, you need a separate sample from each part of the field that differs in cropping history. For example, if one part was planted to grain the previous season while the rest was in cotton, you will need one sample from each area. Before sampling, contact the lab and follow any special instructions. Generally, you will need to follow these steps in each area with a uniform cropping history:

- Use a shovel or soil tube to collect soil to a depth of 1 foot (30 cm) in at least three randomly chosen places. Place the soil in a bucket or bag.
- When you have collected all the soil from one area, mix it thoroughly, then transfer about 2 to 4 ounces (50 to 100 g) of soil to a plastic bag or other moisture-proof container.
- Label each sample with a field number or other identification.
- Keep the samples cool. Don't leave them in the sun or in a vehicle where they will get hot. Send or deliver them to the lab as soon as possible.

Tolerant Varieties. No true resistance to Verticillium wilt is available in current cotton varieties, but some are quite tolerant compared to others. Current Pima varieties are very tolerant. Tolerant upland varieties include Acala SJ-5 and SJC-1 grown in the San Joaquin Valley and the following Acala varieties grown in New Mexico: 1517-75, 1517-77BR, 1517-El, and 1517-E2. Tolerance in upland cottons is based on a complex of genes that enables plants to produce fruit under adverse conditions. It does not prevent infection but appears to restrict the movement of the fungus within the plant. The mechanism of tolerance in Pima cottons has not been established.

In soils where inoculum density is high enough to cause yield loss in one variety, it is often possible to increase yield by planting a more tolerant variety. For example, Acala SJ-2 has produced the best yields in most of the San Joaquin Valley for several years, but Verticillium wilt has reduced yields in some fields where SJ-2 has been planted continually. Where the inoculum level has increased to about 10 microsclerotia per gram, yields have generally increased significantly following a change to SJ-5. Of course, the yield of a new variety depends on the nature of the soil and on other local conditions as well as on the level of Verticillium inoculum.

Changing varieties will not, however, prevent yield loss indefinitely. Inoculum density will increase as long as cotton is planted without rotation, since *Verticillium* infects both tolerant and susceptible plants and produces new inoculum even if it does not cause foliar symptoms or

Symptoms of Verticillium wilt on leaves begin as patchy yellowing between the main veins.

Plants in advanced stages of Verticillium wilt may have most or all of the leaves affected. Injured leaves may drop or may remain on the plant.

Brown streaks appear in the xylem of plants with Verticillium wilt; this discoloration also occurs in plants with Fusarium wilt.

reduce yields. Also, continual planting of a single tolerant variety may favor more virulent strains, which may gradually increase as a proportion of the total inoculum. In any case, the presence of a high level of inoculum in the soil means that there is a risk of severe losses in seasons when low summer temperatures favor Verticillium wilt.

Crop Rotation. Rotation with crops not susceptible to Verticillium wilt is the best way to keep the level of inoculum in the soil from increasing. Rotation reduces the inoculum level because the inoculum gradually decays and is not replaced unless a susceptible crop is planted. The best rotation crops are grass family crops such as corn, small grains, rice, and sorghum, although alfalfa and sugarbeets are also useful. Fallowing can also reduce inoculum density if the field is kept free of broadleaved weeds susceptible to Verticillium wilt.

In San Joaquin Valley soils where the initial level of inoculum is low, rotations in which cotton is planted no more than 2 years out of 5 and which include at least one season in a grass family crop have kept *Verticillium* from increasing to damaging levels. In New Mexico, a rotation including cotton, alfalfa, and barley has significantly reduced losses from Verticillium wilt. Yellow sweetclover and hubam have also been useful rotation crops.

Once the *Verticillium* inoculum has reached a high level, it may be necessary to keep cotton and other susceptible crops out of the field for several years to reduce it significantly. It is difficult to predict how long a rotation must last to achieve a significant reduction, although experience with local soils having a similar cropping and disease history can serve as a guide. Take soil samples each year to monitor the decline in inoculum level, and wait until it reaches a low level before replanting cotton.

Paddy rice has special value as a rotation crop, although it can be planted only in certain soils. A season of paddy rice culture reduces *Verticillium* inoculum density much more than plantings of other rotation crops, greatly reducing the time needed to bring a high density down to a level safe for cotton. The long period of summer flooding required for paddy rice culture apparently contributes to the breakdown of the inoculum.

Plant Density. Yield loss due to Verticillium wilt is generally less in narrow-row plantings than in standard rows in the same soil. The reason is that plants grown under the high density conditions used in narrow-row culture have a relatively small root volume. The smaller a plant's root volume is, the less chance there is that its roots will encounter the Verticillium fungus and become infected. Also, there is a greater chance for healthy plants in dense stands to compensate for those that are diseased. Although narrow-row culture is not currently practiced where cotton is spindle picked, it can be used where cotton is stripper harvested, as in eastern New Mexico. Increasing plant density in standard rows has little effect on losses due to Verticillium wilt.

Fertilizer and Water. Plants deficient in potassium are more susceptible to Verticillium wilt. Routine preplant soil analysis should include an assay for potassium; add potassium at recommended rates if it is too low. Read the section in this chapter on potassium deficiency before adding potassium to soil.

Don't apply excessive nitrogen or water to fields with Verticillium wilt, since they may encourage unnecessary vegetative growth and prolong the fruiting period. A certain amount of water stress can reduce the impact of Verticillium wilt, but the yield reduction caused by the stress is usually enough to offset any positive effect.

Solarization. Solarization—the heating of soil under clear plastic tarps—has been tested successfully for control of Verticillium wilt in the San Joaquin Valley. Most Verticillium inoculum is within 1 foot (30 cm) of the surface, and heat trapped by the plastic raises the soil temperature enough at that depth to kill most of it. Solarization also kills other pathogens, including fungi that cause seedling disease, as well as some insects and many weed seeds. One treatment controls *Verticillium* for about 2 years. For best results, soil moisture must be near field capacity. Either lay the plastic over preirrigated soil or apply water under the plastic by running it into the wheel furrows created when laying the plastic. Because solarization must be done in summer and takes several weeks, it requires rotation with a winter crop. A bulletin with more information on this method is listed in the References.

Fusarium Wilt/ Nematode Complex
Fusarium oxysporum f. sp. *vasinfectum* and *Meloidogyne incognita*

Fusarium wilt is a major disease of cotton in the southeastern U.S., but in the Western Region it is known only in the San Joaquin Valley, where it forms a disease complex with root-knot nematodes. So far, the Fusarium/ nematode complex has been confined to certain parts of the southern San Joaquin Valley. However, it is possible that it could become more widespread if reduced availablity of soil fumigants results in a general increase in root-knot nematode populations in sandy soils.

The complex may begin at any stage of plant growth, but the earliest symptoms usually appear on lower leaves at first flowering. Small areas, usually near the margins, first turn yellow, then brown. These areas gradually expand toward the center of the leaf until the entire leaf dies and drops. Symptoms then appear on leaves further up

Plants with Verticillium or Fusarium wilt show xylem discoloration in petioles and veins as well as in stems.

Symptoms of Fusarium wilt on leaves usually begin at the edge and advance toward the center, creating a wedge-shaped pattern.

Plants diseased due to the Fusarium wilt/nematode complex always have root galls and usually have some of the lateral roots blackened.

The Fusarium wilt/nematode complex can kill plants early in the season, leaving bare spots.

the plant. Plants infected early are often stunted and may die before maturing any bolls. Losses have been severe in some fields.

The xylem of infected plants turns dark brown or black. If you cut a cross section of stem with a knife, you will find a dark ring just inside the bark. This discoloration may eventually extend to the center of the stems and upward throughout the plant. Quite often, one or two of the lateral roots of plants damaged by the Fusarium wilt/nematode complex are blackened. These roots may be the sites where infection began.

Fusarium wilt symptoms can be confused with those of Verticillium wilt, and it is not always possible to distinguish the two in the field. The differences are discussed in the section on Verticillium.

The fungus survives for long periods in soil and on debris from infected plants. Roots become infected when growing root tips encounter the fungus. Root-knot nematodes cause galling and physiological changes which predispose plants to infection by *Fusarium*. Even cotton varieties otherwise resistant to *Fusarium* lose their resistance when infected by root-knot nematodes. Once the fungus has invaded the vascular system, the disease develops in a manner similar to that described for Verticillium wilt.

Acala SJ-5 and SJC-1 are more tolerant to the Fusarium/nematode complex than Acala SJ-2 in San Joaquin Valley soils. Where the disease has appeared in Acala SJ-2, a change of variety may improve yield. Otherwise, management is based on monitoring and control of root-knot nematodes. Recent evidence shows that the Fusarium wilt fungus is present on gin trash and seed from diseased plants; use of infected seed or infested gin trash may favor increased incidence and spread of the disease.

Phymatotrichum Root Rot
Phymatotrichum omnivorum

Phymatotrichum root rot, also known as cotton root rot and Texas root rot, occurs in alkaline soils with low organic matter in Arizona, New Mexico, Texas, and Mexico. In California, the disease occurs only in the Colorado River basin, although the fungus has also been found in small areas in the Imperial Valley. The fungus causes severe losses in cotton, alfalfa, stone fruits, grapes, sugarbeets, and many other crops and ornamentals; it has more than 2,000 hosts, all broadleaved plants. No reliable, economically feasible control measure is now available.

Symptoms and Damage

Phymatotrichum root rot can begin at any time in the season, but aboveground symptoms do not appear until the soil gets very warm—usually in mid-July or later. Infected plants wilt and die within a few days. By the time plants wilt, the outer layer of root tissue, the cortex, is entirely dead and easily sloughs off. Roots may be entirely decayed a foot or so below the surface. The roots are covered with a network of brown fungal strands that are readily visible with a hand lens. Plants killed by cotton root rot do not drop their leaves.

Root rot usually occurs in irregular circles that may persist for many years. The circles do not expand and the fungus does not spread in irrigation water or in cultivation. In nonirrigated areas, root rot may appear to subside in years with little rainfall, when plants grow poorly and are not as suitable as hosts.

The amount of yield loss depends on the size of the affected area and on the time when plants are killed; the earlier plants die, the lower the yield. Losses are especially severe in Pima cotton.

Seasonal Development

The root rot fungus occurs naturally in many soils in the Southwest, including soils of all textures from fine to sandy. However, it is most common in fine-textured soils along river beds and in valleys, and is less common in mesa soils; it is most damaging in alkaline soils, especially when summer temperatures are high. The fungus grows on native plants such as mesquite, which show no symptoms, and the disease can be severe in newly cleared land.

The fungus grows through soil and on roots in the form of strands that consist of numerous intertwined filaments or hyphae. Under a hand lens, the strands look like braided brown rope. They have numerous projecting, cross-shaped hyphae; these hyphae are visible under a microscope at 80 to 100x and are a diagnostic feature of the fungus. At certain times, some filaments form sclerotia—resistant bodies that remain viable in soil for years when growth conditions are unfavorable.

The fungus may also form spore mats on moist soil around diseased plants. The mats are white to light brown, up to 16 inches (40 cm) in diameter and up to ¾ inch (2 cm) thick. The role of the spores in the disease cycle is unknown. The presence of spore mats alone does not necessarily mean the root-rot fungus is present, since other fungi may form similar mats. Spore mats must be examined in the laboratory for accurate identification.

Most infections in cotton originate from sclerotia deep in the soil. The sclerotia germinate when conditions are favorable, producing strands capable of growing for a considerable distance through the soil. When a strand of the fungus contacts a root, it proliferates over the surface and branches into a network that penetrates and destroys the cortex. The fungus spreads over the surface of roots toward the crown and may also spread to adjacent plants by growing along laterals. When the host plant dies, the strands grow back down through the soil and produce new sclerotia that remain as a source of infection in later seasons.

Management

Crop rotation, green manuring, and deep tillage can reduce losses due to root rot, although there is no control measure effective in all cases. Grass family crops such as corn, sorghum, and small grains are not susceptible to root rot and may be useful as rotation crops in infested soils. Studies in Arizona have shown that rotating to a grass crop results in improved yield when cotton is replanted, although the degree of improvement is not always significant.

Green manuring affects cotton root rot because the root rot fungus is adapted to a soil environment where organic matter content is low. Adding organic matter apparently reduces growth of the fungus by favoring other soil organisms that may inhibit its growth or compete with it for nutrients. Small grains have been used as green manure in New Mexico, but the high cost of growing them limits the practicability of this technique. Animal manure added at rates up to 20 tons per acre may be useful as a spot treatment.

Deep, moldboard plowing reduces root rot by exposing the sclerotia to drying, although the cost of fuel usually makes this technique impractical. For best results, turn the soil to a depth of 2 or 3 feet, then leave the field fallow during the summer.

Researchers are testing fungicides for use against cotton root rot, but none is now registered. Soil fumigants and sterilants have not proved effective. There are no cotton varieties resistant to root rot.

Boll Rots

Boll rots are most prevalent in hot, humid weather in the low desert areas, although some infected bolls can be found in almost any cotton field in late summer. Losses are greatest in rank stands, in cotton exposed to rain or long periods of high humidity during boll maturation, and in fields sprinkler irrigated while bolls are opening. These rots mainly affect bolls on the lower half of plants.

Most boll rots in the Western Region are caused by fungi, mainly species of *Aspergillus*, *Rhizopus*, and *Fusarium*. *A. flavus* causes a special problem in the low deserts due to aflatoxin contamination (see below). In the San Joaquin Valley, the most common rot organism is *Rhizopus nigricans*, which commonly affects 5 to 10% of the bolls in rank fields. A rot caused by the fungus, *Nigrospora oryzae*, also occurs there regularly, but at low levels. A bacterium, *Erwinia herbicola*, is a minor boll rot organism in the deserts and the San Joaquin Valley.

Most boll rot organisms cannot penetrate healthy carpels, so infection usually results from airborne spores that land on moist lint after bolls have cracked. The incidence of boll rot is higher where damage due to pink bollworm, tobacco budworm, or other insects provides a route for infection. Spores of *N. oryzae* are carried by a mite, *Siteroptes reniformis*, which is not related to spider mites.

The lint of infected bolls usually turns yellow, brown, or black and fails to fluff normally. Damaged bolls have reduced lint quality, and severely damaged bolls usually cannot be harvested.

Any practices that limit rank growth and reduce the humidity of the canopy help reduce losses from boll rots. Avoid using excess nitrogen and water that favor rank growth. In furrow-irrigated fields, use alternate furrow irrigation when rank stands or hot humid weather favor boll rots. Don't irrigate with sprinklers while bolls are opening in humid weather. Where boll rots are unusually severe, it may be worthwhile to defoliate the lower parts of plants to improve air circulation in the canopy. Follow recommended control measures to prevent excessive insect damage to bolls.

Aflatoxin Contamination
Aspergillus flavus

Aflatoxins are carcinogens produced in cottonseed, nuts, grains, and other foods by the fungus, *Aspergillus flavus*. They cause cancer of the liver in some animals and possibly in man when present in feed or food. The fungus is widespread in the U.S., but significant contamination of cottonseed with aflatoxin is limited to elevations below 2000 feet in western Arizona, southern California, and the lower Rio Grande Valley of Texas. Losses result from the fact that contaminated cottonseed and cottonseed meal cannot be used as feed unless it is mixed with clean feed or treated with ammonia. At present, there is no reliable way to prevent contamination.

Symptoms and Damage

The lint of bolls infected by *A. flavus* is stained yellow. Infection can often be recognized by examining the lint under ultraviolet light, which produces a greenish yellow fluorescence in the discolored areas. Fluorescence of lint is a good sign that the attached seed is contaminated; however, the absence of fluorescence does not necessarily mean there is no contamination, since the fluorescent substance can be washed off by rain or broken down by light before harvest. Another symptom of *A. flavus* boll rot is that seeds are abnormally "bald" or lacking in linters.

Although infected lint is discolored and weakened, the main damage caused by *A. flavus* is the production of

S. J. INGLE/M. R. DAVIS

As shown in this infrared aerial photo, Phymatotrichum root rot usually occurs in round patches that may cover a large part of a field.

Plants with Phymatotrichum root rot die suddenly. Dead leaves remain on the plant.

The root system of plants with Phymatotrichum root rot consists only of the taproot, which has a brown, shredded surface.

The shredded cortex of taproots affected by Phymatotrichum root rot easily sloughs off.

T. E. RUSSELL

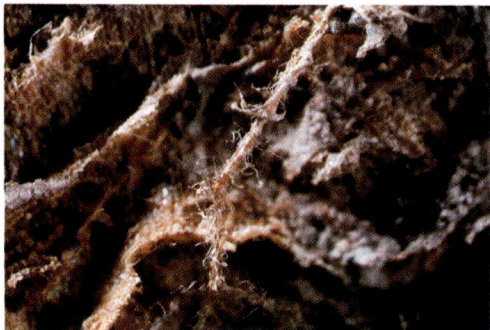

Under a hand lens, the strands formed on the surface of infected roots by the *Phymatotrichum* fungus resemble twine.

Lint infected with boll rot organisms usually turns brown or black and fails to fluff. This boll is infected with the fungus, *Aspergillus niger*.

aflatoxins, which occurs only in the seed. Whole cottonseed and cottonseed meal contaminated with more than a certain level of aflatoxins must be treated with ammonia before it can be used as feed; the extra processing cost represents a financial loss to growers in the affected areas. Cottonseed oil is usually free of contamination, since normal processing leaves nearly all aflatoxins in the meal.

Seasonal Development

A. *flavus* lives on dead plant debris, producing spores that are distributed by wind and insects. The fungus cannot penetrate intact boll walls, so infection occurs only while bolls are opening or when chewing insects such as pink bollworm or *Heliothis* leave holes that allow the spores to reach the moist lint. Under favorable temperature and moisture conditions, the spores germinate and the fungus grows on lint fibers, eventually reaching the seed. Inside the seed, the fungus produces aflatoxins in the course of its normal metabolism.

Although A. *flavus* is widespread, it infects cottonseed only where temperature and humidity are high during the period when moist lint is exposed. Where the nighttime low is consistently below 70° to 75° F (21° to 24° C), as it is in most U.S. cotton-producing areas, the level of contamination remains low. In the low deserts, however, most boll infections occur from early August to mid-September, when nighttime minimums are commonly higher and when humidity is likely to be high. Most bolls opening in this period are in the lower half of the canopy, where humidity is highest.

The moisture content of lint must be above about 15% for the fungus to grow on it. In relatively dry areas, such as the San Joaquin Valley, bolls usually open rapidly after they have split and lint dries quickly. In more humid areas, bolls open slowly after cracking; the lint dries slowly and is exposed to infection for a much longer time.

Insects, especially pink bollworm, contribute to aflatoxin contamination because their damage provides sites where the fungus can enter bolls. Damaged bolls may also open slowly or fail to open completely. The aflatoxin level in bolls damaged by pink bollworm is generally higher than in undamaged bolls, although the proportion of bolls that are contaminated is not strongly associated with the level of pink bollworm infestation. Certain insects, including stink bugs and lygus bugs, carry the fungus in and on their bodies and may play a role in spreading it.

Management

Tolerances set by the U.S. Food and Drug Administration for aflatoxin contamination in feeds range from 20 to 300 parts per billion (ppb), depending on the intended use of the feed. The lowest tolerance applies to feed for

Seeds infected with the aflatoxin fungus have few linters and their lint is stained yellow.

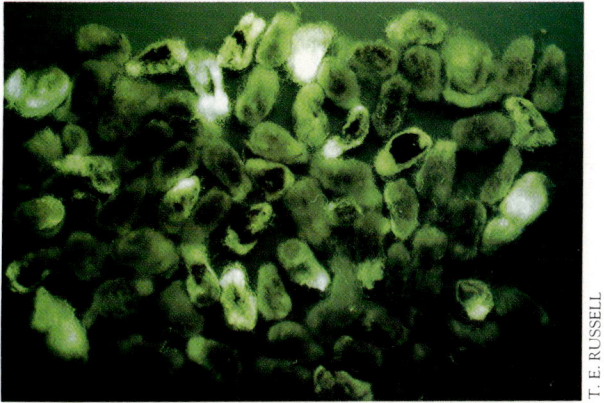

Under ultraviolet light, lint infected with the aflatoxin fungus is fluorescent green.

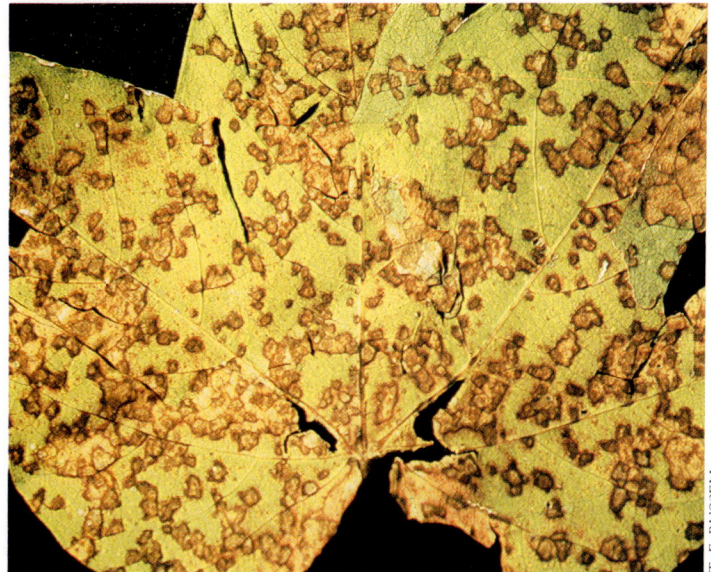

On true leaves, bacterial blight produces brown spots between the smaller veins. Because of the shape of these spots, the disease is also known as angular leaf spot.

dairy cattle. Cottonseed and meal with a level of contamination above the tolerance must either be treated with ammonia or mixed with uncontaminated feeds to produce a feed mixture that is below the tolerance.

Because most of the aflatoxin is in the meal, which comprises about half of the total seed weight, the level of contamination in meal is usually about twice that of the whole seed. For example, the level in whole seed must be 10 ppb or less for the level in processed meal to be less than 20 ppb. There is no way to estimate accurately the amount of contamination in the field, but some commercial laboratories are equipped for sampling cottonseed lots after harvest and analyzing the seed for aflatoxins.

There is no reliable way to prevent aflatoxin contamination when weather conditions favor infection by *A. flavus*, but insect control and practices that improve air circulation in the canopy can help reduce it. Avoid using excessive fertilizer and water that promote rank growth, since the dense canopy in a rank stand favors boll rot fungi, including *A. flavus*. Alternate-furrow irrigation, skip-row planting, the okra leaf characteristic in some cotton varieties, and bottom defoliation can help to reduce humidity in the canopy, but these measures alone will not reduce contamination significantly when other conditions favor the fungus.

You can reduce the level of contamination in the bulk of the seed from a given field by keeping the seed cotton harvested by spindle pickers separate from that harvested by ground gleaners. Spindle pickers harvest relatively few bolls that have tight locks due to insect damage or boll rots, while ground gleaners pick up many damaged bolls. Since most of the aflatoxin is in damaged bolls, the level of contamination in seed cotton harvested by ground gleaners is much higher than it is in seed cotton harvested by spindle pickers. In seasons when the overall level of contamination is low, separating the two lots could provide a large quantity of seed with a low enough level of contamination that the meal could be used as feed.

Bacterial Blight
Xanthomonas campestris pv *malvacearum*

Bacterial blight is a serious disease of cotton in areas such as the Mississippi Delta, where warm, wet weather is common during the growing season. It is of little importance in the Western Region, although it causes occasional losses in eastern New Mexico. Blight reduces yield through defoliation and boll infection, and it lowers grades by spotting lint.

Blight appeared in the San Joaquin Valley in the 1950s, apparently introduced on contaminated seed. It was eradicated by 1962 through a program of plowdown and seed sanitation and has not appeared in California since. State inspection of planting seed is a major factor keeping blight out of the San Joaquin Valley; some contaminated seed lots are destroyed nearly every year. Although normal weather in California is not especially favorable for blight, bacteria introduced on seed could spread rapidly and cause severe losses under sprinkler irrigation.

Blight also occurs in Arizona, mostly at higher elevations, but it usually appears so late in the season that it causes no loss and is seldom noticed. The disease was more common there before the use of acid delinted seed became widespread.

Bacterial blight can affect cotton plants at any stage of growth. On cotyledons and leaves, it first appears as watery, oily-looking lesions that are darker than surrounding tissue. The lesions gradually expand, dry out, and turn brown. Lesions on cotyledons are often round, but lesions on true leaves usually have an irregular, angular shape that gives the foliar phase of the disease the name, "angular leaf spot." Infections may also extend along major veins, turning them dark and leaving a strip of diseased blade tissue along the vein. Infected leaves usually drop. On stems and branches, including stems of seedlings, bacterial blight produces "blackarm"—black cankers that may girdle stems. Lesions on bolls are round and sunken. They are water soaked at first, gradually darkening to brown or black. When the bacteria penetrate the carpel wall, they produce a dark colored rot in the lint.

Blight bacteria are carried on or within seed from infected bolls and in debris from infected plants. Infection occurs when the bacteria are splashed or blown onto plants. Because free water is required for infection, the disease is most common under sprinklers and in areas such as eastern New Mexico where summer storms occur during warm weather. Hail injury is especially damaging, since it provides points of entry for the bacteria.

The use of clean planting seed is the first step in preventing bacterial blight. Since sprinkler irrigation favors blight, it is essential not to use planting seed grown under sprinklers. Acid delinting, although it does not eliminate contamination completely because bacteria may be inside the seed as well as on the surface, greatly reduces the chance of infection from seed. Fungicide seed treatments do not prevent blight.

Resistant varieties suitable for New Mexico, such as Acala 1517–77BR, 1517–El, and 1517–E2, can greatly reduce losses in that area. Stripper cottons planted in New Mexico are susceptible, as are most varieties planted elsewhere in the Western Region.

In fields where bacterial blight has appeared, you can reduce or eliminate it by shredding the stalks thoroughly, then discing to a depth of 12 inches (30 cm) or more and replanting the following season with clean seed. If possible, plant an alternate crop before replanting cotton. In areas with little winter rain, a fall irrigation, preferably applied with sprinklers, is needed after discing to promote decomposition of infected debris. Avoid using sprinklers after cotton has emerged if blight occurred the previous season or if you suspect that infection is likely.

Southwestern Cotton Rust
Puccinia cacabata

Southwestern cotton rust is a fungal disease associated with summer rains in New Mexico, southern Arizona, western Texas, and northern Mexico. Usually a minor problem, it occasionally causes losses in limited areas when rainy weather favors the fungus.

Small, yellow to orange spots caused by rust usually appear on leaves about a week after a rain. They may also appear on bracts, young bolls, and stems, enlarging to about the size of a dime and gradually darkening to brown or purple. Those on lower leaf surfaces develop protruding clusters of yellow, spore-bearing bodies. Infected leaves may drop and lesions may girdle stems or cause them to break in the wind. In severe cases, especially when rust occurs together with Alternaria leaf spot, defoliation may be extensive enough to reduce yield and to cause bolls to open prematurely, reducing quality. Lesions on bolls are superficial and seldom damage lint directly, but infections on peduncles—the boll stalks—may cause bolls to break off.

The rust on cotton is one of three phases in the life cycle of the rust fungus. The others occur in fall and winter on native grama grasses, *Bouteloua* spp. The spores that infect cotton are produced by the overwintering stage on grama grasses. Infection occurs when spores carried by wind fall on cotton plants after summer rains. The life cycle of the fungus is completed when spores produced on cotton are blown back to grama grasses.

Spores produced on cotton do not reinfect cotton plants, so each episode of infection is due to a new shower of spores from grama grasses. Damage due to a single spore shower is seldom significant, but if there is a series of showers and if conditions remain humid and mild, repeated infections may cause substantial defoliation, loss of bolls, and breaking or lodging of plants.

Although spores may be carried many miles, damaging infections occur only when infected grama grass is within a quarter mile (400 m) of the field. If there is grama grass near your field, check in spring for the black, raised rust lesions on the stems; infected stems look like they have been splattered with tar. Destroying infected grama grass by plowing, burning, or grazing may reduce the chance of infection in cotton.

Certain fungicides can prevent cotton rust if applied before spores fall on the plants. To prevent the disease entirely, it would be necessary to treat cotton before summer rains. Because the disease appears sporadically, however, it is more economical to wait until spots appear on leaves and then treat to prevent further infections if rain and humid weather are expected to continue. Apply the fungicide every 2 weeks until dry weather returns or until harvest. Resistance to rust is available in some Acala breeding lines, but all commercial varieties are susceptible.

Alternaria Leaf Spot
Alternaria macrospora

Alternaria leaf spot is common following late summer storms in New Mexico and at elevations above 3000 feet in Arizona. The disease also occurs elsewhere in the Western Region, but it is usually so minor and/or occurs so late in the season that it is seldom noticed.

Alternaria produces round, dull brown spots on leaves and occasionally on bolls; the spots are very small at first, but may enlarge to as much as ½ inch (1 cm) in diameter. Older spots have dead tissue at the centers which may crack and fall out, causing a shot-hole appearance. Symptoms are most common on lower leaves. Infected leaves may drop, but the limited defoliation is usually more beneficial than harmful, since it occurs late in the season and tends to improve air circulation in the canopy, reducing the chance of boll rots.

Injury to leaves caused by leaf spot is seldom if ever economically significant. Symptoms are more noticeable on Pima cottons, which are much more susceptible, than on upland varieties. Fungicides are available for Alternaria leaf spot, but control is rarely necessary.

Leaf Crumple

Leaf crumple is caused by a virus transmitted by the sweetpotato whitefly, *Bemisia tabaci* (page 83). The disease occurs mostly in the low deserts. Leaves of infected plants are smaller than normal, have yellow mottling between the veins, and are often wrinkled and cupped. Petals may be short and crinkled and may have small bumps on the outer surface. Bracts may also be distorted, and bolls may have small bumps on the surface. The reduced size and number of bolls on infected plants can cause significant yield reduction.

Leaf crumple appears earliest and causes the greatest damage in stub cotton, apparently because the rootstocks provide an overwintering site for the virus. The disease readily spreads to planted cotton when the whitefly vector is abundant. Whitefly populations are highest following mild winters that also favor stub and volunteer cotton, and insecticides applied to cotton for other pests may also contribute to outbreaks. Elimination of stub cotton has greatly reduced the incidence of leaf crumple in the past and is probably the best preventive measure. However,

Southwestern cotton rust produces round, raised spots on leaves.

Alternaria leaf spot causes round spots of dead tissue which may later fall out, leaving a hole.

Leaves affected by leaf crumple are rough and distorted.

A leaf discolored due to potassium deficiency.

Ozone, a component of polluted air, speeds up the aging of leaves and causes yellow to red discoloration where tissue eventually dies.

there is evidence that the virus may overwinter in weeds such as cheeseweed as well as in cotton, so it is possible that the disease could appear whenever large populations of sweetpotato whitefly are present, even in the absence of stub cotton.

Potassium Deficiency

Symptoms of potassium deficiency usually appear relatively late in the season, after plants have set some fruit and when some yield loss may already have occurred. Deficient leaves appear bronze at first, then develop yellow to orange areas between the veins; these areas may die and turn dark. The edges of leaves also may appear burned, and leaves may be crinkled, cupped, and abnormally thick. Symptoms usually appear on older leaves first. Severe deficiency causes defoliation and shedding of squares or young bolls.

A condition that produces potassium deficiency even on soils not low in potassium is widespread in the San Joaquin Valley. It differs from typical potassium deficiency in that it often occurs uniformly throughout a field, while a typical deficiency usually appears only in patches. Also, deficiency symptoms caused by this condition may appear earlier in the season and on younger leaves. The cause of this type of potassium deficiency is unknown, but experimental work suggests that a pathogen, probably a soilborne fungus, may be involved. The deficiency condition found in the San Joaquin Valley often occurs together with Verticillium wilt, and their symptoms can be confused. The effects of Verticillium wilt are worse where the deficiency also occurs.

Unlike typical potassium deficiency, the condition found in the San Joaquin Valley is not eliminated by adding potassium to soil in normal amounts, although yields have been improved significantly by adding amounts in the range of 250 pounds per acre. Yield losses due to

R. B. HINE

potassium deficiency are greater in Acala SJ–2 than in SJ–5 or SJC–1. Where the yield of SJ–2 has been reduced by a deficiency that does not respond to normal potassium rates, you may be able to improve yield by changing to a more tolerant variety.

Follow the guidelines on page 28 to make sure your soil is not actually low in potassium. Because potassium deficiency is worse in compacted soils, you can reduce losses by avoiding unnecessary wheel traffic while the soil is wet and by ripping to break up compacted layers.

Toxicities

Cotton plants may be injured by exposure to toxic substances, including improperly applied pesticides. In the Western Region, the most common toxicities are due to air pollutants and herbicides.

Air Pollutants

Certain gases found in polluted air can injure cotton plants and reduce yields. The most common is ozone, a form of oxygen produced in sunlight from the components of automobile exhaust and other combustion by-products. Ozone kills leaf tissue, reduces photosynthesis, and accelerates the aging process in leaves. Symptoms first appear on the upper surfaces of older leaves. Leaf tissue between the small veins turns yellow and may later become stippled with numerous small purple or brown spots. The leaf's entire upper surface may eventually turn bronze and it may drop.

Ozone injury is widespread on cotton in the San Joaquin Valley, and many growers regard the symptoms as normal. However, studies comparing cotton growth in polluted air with growth in filtered air have shown that concentrations of ozone commonly found in polluted air can reduce the yield of Acala SJ–2 cotton by 15 to 20%. Losses are due mostly to a reduction in the number of bolls. Limited experimental evidence suggests that Acala SJ–5 may be somewhat more tolerant of ozone than SJ–2.

The concentration of ozone and other air pollutants is affected by weather, local patterns of air circulation, and location relative to sources of pollution. Ozone affects large areas but it tends to be most concentrated downwind of urban centers. Sulfur dioxide, a gas produced in nonferrous smelters as well as in the combustion of fossil fuels, causes the same kind of injury as ozone, but the injury is usually limited to fields that are close to smelters or other large, fixed sources of air pollution.

Herbicides

Symptoms of herbicide injury may resemble those caused by pests or nutrient deficiencies. To identify herbicide injury, you may need to have a laboratory analyze

T. E. RUSSELL

Leaves on cotton plants exposed to 2,4–D fail to expand normally, becoming straplike and twisted.

Postemergence herbicides used as directed sprays may cause yellow spots where herbicide contacts lower leaves. Diuron, fluometuron, cyanazine, and prometryn can all cause these symptoms.

soil, water, or plant tissue for residues, but for analysis to be useful, you must tell the lab what to look for. This means that you must know the history of herbicide use in the fields where the symptoms occur and in nearby fields where spray drift or residue in water may have originated. One limitation of laboratory analysis is that the presence of an herbicide residue does not necessarily prove that a certain injury was caused by the herbicide. Farm advisors and professional consultants can help in identifying these problems.

Dinitroaniline herbicides such as trifluralin, pendimethalin, and fluchloralin may inhibit lateral root development and cause stunting if applied at too high a rate or if cotton seed is planted too deep in treated soil (Figure 50, page 120). During cool weather, they may inhibit emergence and slow growth, increasing losses due to seedling diseases.

Substituted ureas such as diuron and fluometuron and triazines such as cyanazine and prometryn can stunt or kill seedlings if applied improperly or at excessive rates. They may also cause yellow or white spots on leaves of mature plants if spray contacts the foliage.

The arsenical herbicides, DSMA and MSMA, can turn foliage red and burn the edges of leaves where spray collects. Exposure usually results from misdirected applications or from intentional application over the tops of small plants. Injury may delay maturity and reduce yield, especially if it is followed by dry weather stress.

Cotton is very sensitive to phenoxy herbicides such as 2,4–D. Exposure to very small concentrations causes severe distortion of leaves and may cause enlargement of stems and tap roots. Injury is usually due to drift from applications in other crops, but it can also result from residues left in sprayers or other equipment used in cotton fields.

Avoid herbicide injury by following the directions on herbicide labels and by making sure before planting that the soil and water do not contain harmful residues.

Weeds

Weeds reduce yields by competing with cotton for light, water, nutrients, and space. Early competition causes the greatest yield reduction, as it slows establishment of the crop and may delay maturity. Late-season infestations interfere with defoliation and harvest and may lower lint grade if the lint is stained or contains excessive weed trash. Weeds, including those growing outside the planted area, may affect the crop indirectly by serving as alternate hosts for insects, mites, or pathogens.

In the Western Region, the most difficult weeds to manage are perennials such as nutsedges, field bindweed, bermudagrass, and johnsongrass, and annuals resistant to preplant herbicides used in cotton. Resistant annuals include nightshades, groundcherries, annual morningglories, and cocklebur; several other species are important locally.

Monitoring

To make the best choice of herbicides and rotation crops, conduct weed surveys regularly to find out which weeds are present and how their populations are changing. Adjacent fields often have very different weed populations due to differences in cropping history or soil type, so you must survey each field separately. Because many herbicides are effective only on germinating weeds, you need to know which species are present *before* they grow; keeping accurate survey records will provide this information.

To conduct a weed survey, walk through the field in a random pattern and rate the degree of infestation for each species. Use a numerical scale (Figure 44) or rate infestations as light, medium, or heavy. Check fencerows, turnrows, and ditchbanks as well as the field itself. Pay special attention to perennials; sketch a map of the field and mark where they occur. Ask a farm advisor or county agent for help if you find weeds you can't identify.

Since different annual weeds are present at different times of year, you will need to check each field at least twice a year. Conduct your first survey in spring before preparing the field, so you will see which winter annuals are present. This information will help in planning winter crops and fall or winter herbicide applications. Survey

again during the growing season to check the effectiveness of control measures and to look for summer annuals.

Crop Rotation

Whenever you plant the same crop year after year, there is a good chance that one or more weed species will increase because they are adapted to the same conditions as the crop. Also, repeated use of the same or similar herbicides usually favors certain weeds that can tolerate them. Where cotton is planted without rotation and

WEED INFESTATION RECORD

Field _____ Current Crop _____ Date Planted _____

Herbicide _____ Rate _____ Date Applied _____

Previous Crop_____ Date of this survey_____

BROADLEAF ANNUALS						ANNUAL GRASSES					
cheeseweed	1	2	3	4	5	barley (volunteer)	1	2	3	4	5
cocklebur	1	2	3	4	5	barnyardgrass	1	2	3	4	5
cudweed	1	2	3	4	5	cupgrass	1	2	3	4	5
groundcherry	1	2	3	4	5	fingergrass	1	2	3	4	5
horseweed	1	2	3	4	5	junglerice	1	2	3	4	5
Indian tobacco	1	2	3	4	5	large crabgrass	1	2	3	4	5
jimsonweed	1	2	3	4	5	lovegrass	1	2	3	4	5
knotweed	1	2	3	4	5	sandbur	1	2	3	4	5
lambsquarters	1	2	3	4	5	sprangletop	1	2	3	4	5
London rocket	1	2	3	4	5	wild barley	1	2	3	4	5
morningglory	1	2	3	4	5	wild oats	1	2	3	4	5
mustards	1	2	3	4	5	_____	1	2	3	4	5
nettleleaf goosefoot	1	2	3	4	5	_____	1	2	3	4	5
nightshade, black	1	2	3	4	5	_____	1	2	3	4	5
nightshade, hairy	1	2	3	4	5	_____	1	2	3	4	5
pigweeds	1	2	3	4	5	_____	1	2	3	4	5
prickly lettuce	1	2	3	4	5						
puncturevine	1	2	3	4	5						
purslane	1	2	3	4	5	PERENNIALS					
Russian thistle	1	2	3	4	5	alkali sida	1	2	3	4	5
shepherdspurse	1	2	3	4	5	bermudagrass	1	2	3	4	5
sowthistle	1	2	3	4	5	field bindweed	1	2	3	4	5
sunflower	1	2	3	4	5	johnsongrass	1	2	3	4	5
velvetleaf	1	2	3	4	5	nutsedge, purple	1	2	3	4	5
_____	1	2	3	4	5	nutsedge, yellow	1	2	3	4	5
_____	1	2	3	4	5	Russian knapweed	1	2	3	4	5
_____	1	2	3	4	5	silverleaf nightshade	1	2	3	4	5
_____	1	2	3	4	5	tolguacha	1	2	3	4	5
_____	1	2	3	4	5	_____	1	2	3	4	5
_____	1	2	3	4	5	_____	1	2	3	4	5
_____	1	2	3	4	5	_____	1	2	3	4	5

Figure 44. This example of a weed survey sheet lists weeds common in the San Joaquin Valley. To draw up survey forms for other areas, ask a farm advisor or county agent for a list of recognized common names of weeds found locally. Circle the appropriate number to indicate the degree of infestation: 1 = very few weeds; 2 = light; 3 = moderate; 4 = heavy; 5 = very heavy.

where dinitroanilines are applied each year, resistant weeds such as nightshades, annual morningglories, cocklebur, and groundcherries often increase. You can prevent resistant weeds from increasing by rotating to crops that have different growth habits from cotton and in which different cultural practices and herbicides are used. Many annual weeds are kept in check by a rotation program which includes a small grain, alfalfa, and a cultivated crop such as cotton.

Field Selection and Preparation

Weed control is easier in properly prepared fields that are not already infested with perennial weeds or resistant annuals. If the field is infested with problem weeds, consider eliminating them first by planting an alternate crop or by using other methods outlined in the sections on specific weeds.

To do a good job of cultivating and incorporating herbicides, you need a seedbed that is level, firm, and free of large clods and debris. Disc the debris from previous crops or stands of weeds early enough that it will have time to decay before planting. Shred cotton stalks as soon as possible after harvest to promote uniform burial and rapid decay of debris.

Make sure that planting depth, soil moisture, and fertilizer placement are optimum for rapid emergence and growth of cotton seedlings. A vigorous, uniform stand of cotton competes more effectively with early germinating weeds and can prevent establishment of many late-season weeds.

Leveling that eliminates low spots helps to prevent such problem weeds as nutsedge from building up in poorly drained areas. If possible, however, it is best to eliminate small patches of perennials before leveling to reduce the chance of spreading them throughout the field.

Hand Weeding and Sanitation

Hand weeding is often needed when weeds survive herbicide treatments and cultivation. Although expensive, hoeing and hand pulling can prevent substantial losses that would otherwise occur where other methods fail to prevent such weeds as nightshades and annual morningglories from competing with cotton. To reduce hoeing costs, use close cultivation to limit weeds to a narrow strip along the seed row. Mechanical thinning can further reduce hoeing time. Before sending in a hand crew, show workers the weed seedlings you most need to control. Hoeing is inefficient if workers fail to distinguish weeds such as nightshades from cotton.

Don't limit weed control efforts to the cultivated part of the field. Try to keep fencerows and ditchbanks free of weeds, and make a special effort to destroy stands of perennial grasses, nightshades, field bindweed, jimsonweed, and other weeds that will pose difficult problems if they invade the cultivated area.

Rogue out small infestations of problem weeds from within the field. Remove plants that have fruit or seeds from the field and destroy them if possible. Hand removal is time consuming, but it can be a good investment, since seeds from a few plants can later produce dense stands that are expensive to manage.

Cultivation

Schedule cultivation to destroy weeds at the 1- to 2-leaf stage, before they deplete soil moisture and while they are easily dislodged. Larger weeds require deeper cultivation, which makes it hard to maintain bed shape and may damage cotton roots. Cultivating too deeply may also bring up soil and weed seeds from below the layer treated with herbicide, resulting in more weed germination.

With the right selection and arrangement of tools, you can dislodge weeds from furrows and bed tops while burying weeds in the seed row with a layer of soil. Equipment effective for this purpose includes rolling cultivators (ground-driven rotary cultivators) and reversed disc hillers followed by sweeps (Figures 45, 46). The layer of soil moved into the seed row must be shallow while cotton seedlings are small but it can be an inch deep once plants are well established. In some soils and with some equipment, a shield may be needed to keep large clods out of the seed row (Figure 47). Sweeps or knives used in early cultivation can be set as close as 2 inches (5 cm) to the seed row as long as they are shallow.

Another way to destroy small weeds in the seed row is to use row weeders (Figure 48). When set just below the soil line, these tools pinch the soil in the seed row, causing it to buckle and dislodge small weed seedlings. Cotton seedlings must be well enough established that they will not be dislodged.

Keep the tools used in bed shaping and planting carefully aligned so that you will be able to follow the seed row precisely during later cultivation. You can then set cultivating tools close to the seed row without injuring small cotton plants. Guidance devices such as sleds and guide wheels help to keep tools aligned with the beds during cultivation.

Weeder Geese

Geese have been effective for early season weed control in parts of the San Joaquin Valley where grasses and nutsedges are the main weeds in cotton. When properly managed, young geese will feed on grass seedlings and on new shoots of nutsedges and perennial grasses without injuring cotton. Geese about 6 weeks old are best.

Figure 45. Rolling cultivators set up for close cultivation of small cotton.

Figure 46. A set of disc hillers followed by sweeps can dislodge small weed seedlings on beds while burying those in the seed row with a shallow layer of soil.

Stock the field with three to five geese per acre after cotton has emerged but before it reaches the 4-leaf stage. Make sure the stand of small weeds is adequate to support the geese, so they will not eat cotton seedlings. A temporary fence of chicken wire and laths about 3 feet tall will confine geese to the field. Protect them from dogs and coyotes and provide clean water, shade, and a small amount of supplemental feed. Remove geese from the field before applying pesticides, and don't replace them until the reentry interval on the label has passed.

Herbicides

To use herbicides successfully, you must know which weeds are present and you need to consider the soil type, crop rotation sequence, and the timing of cultural operations in each field. Timing of application relative to crop and weed growth is critical in determining which herbicide to use and how to apply it.

The herbicide treatment best for a particular field depends largely on whether the field is infested with weeds not susceptible to dinitroanilines, the most common preplant herbicides used in cotton (Figure 49). If the dominant weeds are annuals readily controlled by dinitroanilines, a preplant treatment combined with careful cultivation often provides adequate weed control. Where resistant annuals are present, you may need to combine a dinitroaniline with another material as a preplant treatment, and you may also need one or more postemergence treatments. Fields infested with perennials may need a foliar-applied herbicide. Where perennials or resistant annuals are abundant, it may be more economical to reduce populations by rotating to a crop in which effective herbicides are available before trying to manage these weeds in cotton.

Always READ THE LABEL before using any herbicide and follow the directions carefully. Use application equipment suited to field conditions and calibrate it before you start. Keep in touch with farm advisors or county agents who specialize in weed control, and consult current recommendations from your state's Extension Service when selecting materials and rates. Recommendation pamphlets and a publication on calibrating herbicide sprayers are listed in the References.

Soil-Applied Residual Herbicides. Soil-applied residual herbicides are absorbed by weed seedlings at germination, killing them before or soon after emergence. The substituted dinitroanilines—fluchloralin, pendimethalin, and trifluralin—are the most widely used soil-applied herbicides in western cotton. Properly applied, they are effective against most annual grasses and many broadleaved annuals, but they do not adequately control nightshades, groundcherries, cocklebur, annual morningglories, or certain other weeds. They suppress the growth of perennials, especially those sprouting from seed, but will not control established perennials at rates used in cotton. Dinitroanilines must be mechanically mixed, or *incorporated*, into the soil at the a depth of 2 to 4 inches (5 to 10 cm). Failure to incorporate them properly is a common cause of poor weed control. Dinitroanilines are usually applied at or before planting, but trifluralin can also be used after crop emergence.

Various soil-applied herbicides are available for control of many annual weeds not susceptible to dinitroanilines. Several, including cyanazine, diuron, fluometuron, and prometryn, have a limited degree of foliar activity, so they are effective on small emerged seedlings as well as on germinating weeds. Under western conditions, they are used mostly for controlling weeds in the seed row after cotton emergence. They generally are not incorporated, but are moved into the soil with irrigation water or rain. Prometryn, however, is also useful as a preplant incorporated treatment in combination with a dinitroaniline.

Incorporation. Incorporation tools include finishing discs, spring-tooth cultivators, rolling cultivators, and power-driven tillers. Herbicide labels often specify which is best. You can use finishing discs prior to bed formation if you adjust tractor speed and arrange disc gangs to mix the top few inches of soil thoroughly. Gauge wheels help keep the depth of incorporation uniform. For best results, cross-disc in such a way that neither pass with the discs is parallel to the rows. You can use spring-tooth cultivators in the same way as discs.

Rolling cultivators are versatile incorporation tools, as you can use them before or after forming beds and before or after preirrigation. They can incorporate herbicide uniformly to a depth of about 2 inches (5 cm) as long as the beds are reasonably free of large clods. To use them after bed shaping, place two or three gangs of spider wheels in tandem over the centers of the beds and arrange them so that the front and rear gangs move soil in opposite directions. Use a forward speed of about 5 mph. Avoid removing treated soil from bed tops at planting.

Power tillers are useful for incorporation in fine-textured soils, where they help to eliminate clods and provide more precise placement of herbicide than other equipment. They are used mostly for incorporating band applications on finished beds. Avoid using power tillers with L-shaped blades in wet soil, since they may create a shallow, compacted layer that will interfere with the growth of cotton seedlings. Use a forward speed of about 3 mph or less; be careful not to remove treated soil from bed tops in planting.

Band Applications. A *band* application covers only the bed or seed row, while a *broadcast* application covers the entire field area. For band treatments, calculate the amount of herbicide needed according to the area actually sprayed,

Figure 47. Shields keep clods out of the seed row during close cultivation with a set of harrows.

Figure 48. The slender, pointed row weeders in the center of this cultivation setup buckle the soil in the seed row, destroying small weed seedlings.

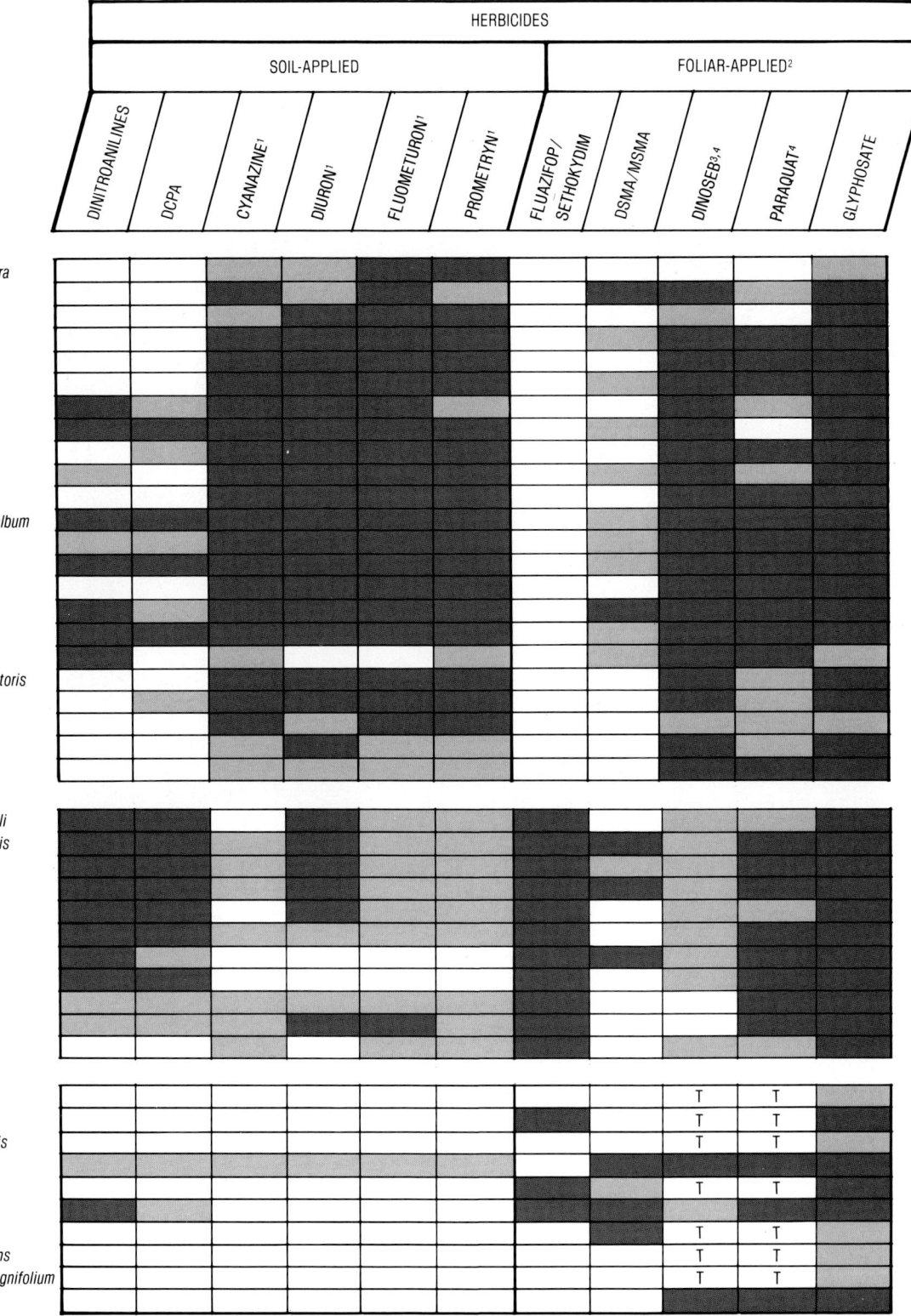

Figure 49. Susceptibility of common weed species to herbicides used in cotton in the Western Region.

not according to the size of the field. For example, if you treat a 100-acre field by spraying a 10-inch band on 40-inch rows, you will be treating only one quarter of the total area, or 25 acres. If the recommended rate is one quart of active ingredient per acre, you would use 25 quarts for the entire 100 acres.

Successful band application requires nozzles such as the 8002E that can apply herbicide evenly over the entire band width. Adjust nozzles carefully and make sure they remain properly aligned with the seed row during application. Remember that the rate of material applied changes according to the placement of the nozzles; a nozzle designed to cover a 20-inch swath will apply twice as much material to the treated area if it is set to spray a band only 10 inches wide. It is usually not practical to apply a band less than about 6 inches (15 cm) wide. For band treatments that will be incorporated, spray a band about 2 inches (5 cm) wider than the strip you plan to cover in incorporation.

Applications in Variable Soils. Application of soil-applied materials is complicated in fields with large sandy streaks or other variations in soil texture, since rates must be adjusted for soil texture. Applying herbicide to sandy soil at a rate intended for finer-textured soil can result in crop injury. One way to deal with sandy streaks is to mark the edges with stakes, then vary the tractor speed during application to adjust the amount of herbicide applied. This method is not practical if you are applying an herbicide as part of another operation, such as planting. If you apply herbicide to a field with sandy streaks but you cannot vary the rate, you may need to accept a certain amount of crop injury in the sandy soil to achieve acceptable weed control in the rest of the field.

Preplant Applications. Residual herbicides applied to the soil shortly before planting protect the crop from weed competition during seedling growth. Preplant treatment in fall or winter can also reduce weed growth on fallow beds, thereby reducing the need for cultivation at planting. Both kinds of preplant treatment are effective mainly against annual weeds. Where perennial weeds are present, you may need special cultivation practices and/or a foliar-applied herbicide after crop emergence in addition to preplant treatments.

Dinitroanilines are widely used for preplant treatments. Several common winter weeds are resistant to dinitroanilines, however, so fields treated in fall or winter may need a foliar-applied herbicide and/or cultivation before planting. Also, slightly higher rates are needed in early applications to compensate for dissipation of the material with time. Oxyfluorfen, another soil-applied residual herbicide, has also been tested on fallow beds and is more effective than dinitroanilines on winter weeds. Other preplant herbicides do not last long enough in soil for fall or early winter application to be effective. Prometryn has been useful as a preplant treatment for nightshades in the San Joaquin Valley, but its residual life is limited so application is not recommended before February 15.

To apply an herbicide before forming beds, use a broadcast application and incorporate immediately, using the technique indicated on the label. After listing, you can apply herbicide either as a broadcast or band treatment. Whenever you apply a dinitroaniline before planting, adjust the planting depth to keep the cotton seed at or near the bottom of the treated soil layer (Figure 50).

In sandy soils, you can apply a band of herbicide during planting by inserting two or three rolling cultivator gangs between the dirt pusher and the planter. Arrange nozzles to spray the soil behind the dirt pusher, where the herbicide will be incorporated by the cultivator gangs. This is a good way to apply a combination of prometryn with a dinitroaniline for nightshade control. To ensure precise alignment of planters, nozzles, and cultivators, use sled-mounted equipment or place gauge wheels on the planter units.

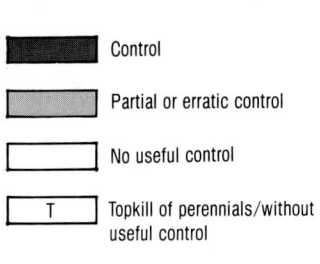

Control

Partial or erratic control

No useful control

T   Topkill of perennials/without useful control

[1] Ratings of cyanazine, diuron, fluometuron, and prometryn are based on soil activity. Effectiveness from foliar activity on emerged seedlings is generally less.

[2] Except as noted, ratings of foliar-applied herbicides on perennials refer to the effects on plants growing from rhizomes, stolons, or tubers.

[3] Effectiveness of dinoseb on grasses assumes application in oil.

[4] Dinoseb and paraquat are restricted pesticides. Obtain a permit from agricultural commissioner or other local authority before purchase or use.

Most of these weeds are illustrated in the *Growers' Weed Identification Handbook;* see references, page 140.

Ratings in this chart are based on a consensus of several researchers. They assume that herbicides are applied at label rates, at the appropriate time relative to weed and crop growth, and with suitable equipment in good condition. Ratings are intended as a guide to effectiveness under average conditions in the Western Region; effectiveness in specific fields may vary according to soil texture, moisture, and other conditions.

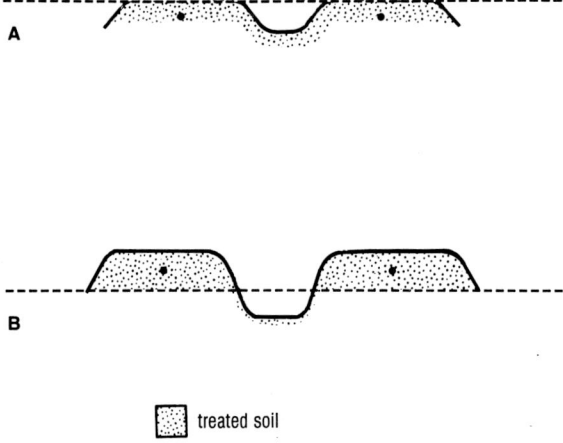

treated soil

Figure 50. When planting into soil that has been treated with a dinitroaniline herbicide, place the seed at or near the bottom of the treated layer (A). If you leave too much treated soil on the beds or plant too shallow (B), the taproot will have to grow through too much treated soil and growth may be retarded.

Tractor-drawn equipment provides the most uniform application of preplant herbicides, but aerial application is permitted for some materials and it may be more convenient on large parcels. If you use an aerial application, make sure the strips treated in each pass of the plane do not overlap. Otherwise, the doubled rate in the overlapping area may injure cotton seedlings or rotation crops. Never apply herbicides by air when conditions favor drift.

Preemergence Applications. Preemergence herbicides are applied between planting and crop emergence. Usually, the herbicide is sprayed in a band over the seed row during planting or shortly afterward. Rainfall or irrigation is then needed to move the herbicide into the soil. Preemergence applications are most common in the desert valleys, where cotton is often irrigated up rather than preirrigated. DCPA, a soil-applied herbicide that requires shallow incorporation followed by irrigation, can be useful as a preemergence treatment. It is effective for most of the same weeds susceptible to dinitroanilines. Use of preemergence herbicides is limited outside the desert valleys, partly because irrigating up tends to favor seedling diseases by cooling the soil. However, preemergence application of prometryn followed by sprinkler irrigation can be worthwhile for heavy infestations of weeds such as nightshades that are not controlled by dinitroanilines.

Postemergence Applications. Postemergence treatments are those applied after the crop has emerged. While cotton plants are less than about 1 foot (30 cm) tall, directed sprays of cyanazine, diuron, fluometuron, or prometryn can destroy small seedlings of groundcherries and other weeds that come up in the seed row after the crop. You can often achieve better control by combining low rates of these herbicides with MSMA. To prevent crop injury, arrange nozzles carefully so that the spray covers the seed row and the lower stem of the cotton plants. Cotton may be injured if the herbicide contacts too much of the foliage, especially if it is sprayed on the growing terminal. Always follow label precautions when using these herbicides. For each one, there is a specific minimum size that cotton plants must reach before they can be treated safely.

Fluometuron can be sprayed over the top of small cotton plants as a salvage operation in cases where numerous weed seedlings escape preplant treatments. Over-the-top treatment delays crop maturity, but a single treatment usually does not affect yield unless the season is too short for the crop to make up for the delay. Injury may be greater in sandy soils, especially if prometryn was applied before planting or if there is an irrigation or rain soon after application.

Herbicide applications after the crop is well established, especially those at the time of the last cultivation, are often called "layby" applications. They are intended to prevent weeds from germinating when it is no longer possible to destroy them in cultivation. Layby applications are usually not necessary in fields with a good stand of vigorous cotton, as long as cultivation is continued until the crop closes over the rows.

Layby herbicides are usually applied as a directed spray that covers the furrow as well as the beds and the base of the plants. Some injury to lower cotton leaves often results, but it does not affect yield as long as the plants are about 2 feet (60 cm) tall or larger at the time of application. Depending on the situation and the material used, shallow incorporation may be needed for best results. Consult the herbicide label for directions. Always irrigate as soon as possible after application. If you cultivate after application, keep the cultivation shallow to avoid removing treated soil from the beds. Don't use equipment that will expose untreated soil in the furrows or throw untreated soil onto the beds.

Foliar-Applied Herbicides. Foliar-applied herbicides are sprayed on the foliage of emerged weeds and are absorbed through leaves and other green tissues. They are generally the most effective materials for control of perennial weeds. Some foliar-applied herbicides will injure cotton, so you may need a shield, hood, or other special equipment to apply them after cotton has emerged. To achieve good control with them, make sure weeds are growing vigorously and are in the right stage of growth as indicated on the label. Treating weeds while they are stressed for water usually results in poor control. The performance of certain foliar-applied materials can be improved by adding oils or surfactants that enhance the ability of the herbicides to penetrate plant tissues.

Paraquat, dinoseb, and glyphosate are foliar-applied herbicides that can be used before planting to kill winter weeds growing on beds listed in fall or winter. Dinoseb is effective on most broadleaved weeds 2 inches (5 cm) tall or less, and it can be effective on many grasses if applied in oil. Paraquat is more effective than dinoseb on grasses, but less effective on certain broadleaved weeds. The two herbicides can be used in combination to kill mixed stands of weeds. Both will burn back the foliage of perennials, but neither provides useful control of perennials. Glyphosate is effective against a broad spectrum of weeds.

Selective grass herbicides (fluazifop, sethoxydim, etc.) are effective against most grasses, including perennials such as johnsongrass and bermudagrass, but they are not effective for broadleaved weeds. They are the easiest foliar-applied herbicides to use after cotton emergence, because they are very safe on cotton. Nonselective foliar-applied herbicides can also be used after cotton emergence if you take appropriate precautions to protect the crop. The methanearsonate herbicides, DSMA and MSMA, can be used to control yellow nutsedge, and glyphosate can be used as a spot treatment for isolated infestations of perennials before boll opening. These special applications are discussed in the sections on specific weeds.

Foliar-applied herbicides, especially glyphosate, are also valuable for destroying weeds outside the field.

Crop Safety and Residues. The effectiveness and safety of soil-applied herbicides varies according to soil texture, organic matter content, acidity (pH), salinity, and other conditions. Make sure you know the soil conditions in each field so that you can make the best choice of materials and rates. Standard soil tests can measure most of the important factors. If you are not certain that an herbicide is safe for cotton under your conditions, test it in a limited area for a year or two before applying it to a whole field. It is a good idea to leave a small untreated area in two places in each field when using residual herbicides. These areas serve as a check to help you determine whether any subsequent crop injury may be due to the herbicide.

Even with materials generally considered safe for cotton, improper application or excessive rates can injure the crop. Common sources of error include improperly calibrated sprayers, incorporating herbicides too deeply, planting too shallowly in treated soil, and overlapping swaths in aerial applications. Another mistake is to apply the same or similar herbicides more than once at rates that add up to more than the total recommended rate per season. For example, if you use trifluralin after crop emergence in a field where it was also applied before planting, make sure the total amount used in the two applica-

tions does not add up to more than the maximum indicated on the label for the whole season.

Some herbicides used in cotton can persist in the soil and may injure the next crop. If you plant to plant another crop following cotton, check the labels of the herbicides used in cotton to make sure they are safe for the rotation crop (Table 11).

The length of time that an herbicide remains in the soil depends on the herbicide chemistry and on such factors as the activity of soil organisms, moisture, temperature, soil chemistry, and tillage. Under arid conditions common in the West, herbicides break down slowly in fallow soil. Recommended herbicide rates are set at levels that minimize soil residues under normal conditions, but reduced irrigation or tillage can increase the persistence of residues. For example, in fields where an herbicide is applied as a broadcast treatment, excessive residues may remain in skip rows or in alternate furrows that are not irrigated.

Table 11. Names, Application Types, and Soil Residue Hazards of Cotton Herbicides.

Common Name	Trade Name(s)	APPLICATION Soil-applied Residual	Foliar	Soil Residue Harmful to Rotation Crops[1]
cyanazine	Bladex	•	•	
DCPA	Dacthal	•		
dinoseb	various		•	
diuron	Karmex and others	•	•	•
DSMA	various		•	
fluazifop	Fusilade		•	
fluchloralin	Basalin	•		•
fluometuron	Cotoran	•	•	•
glyphosate	Roundup		•	
MSMA	various		•	
oxyfluorfen	Goal	•	•	
paraquat	Gramoxone, Paraquat		•	
pendimethalin	Prowl	•		•
prometryn	Caparol	•	•	•
sethoxydim	Poast		•	
trifluralin	Treflan	•		•

1. See herbicide labels for required intervals between application and following crops.

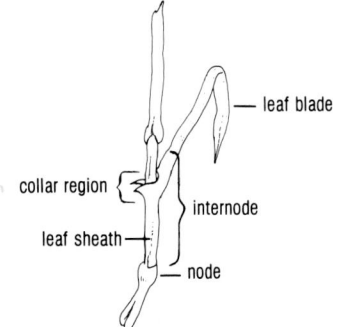

GRASS STEM AND LEAF

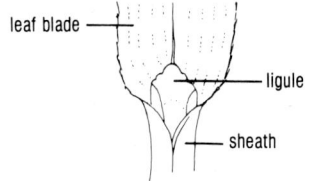

COLLAR REGION OF GRASS

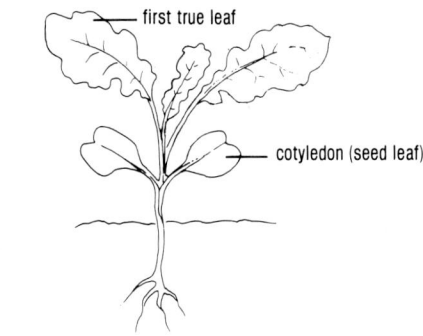

BROADLEAF SEEDLING

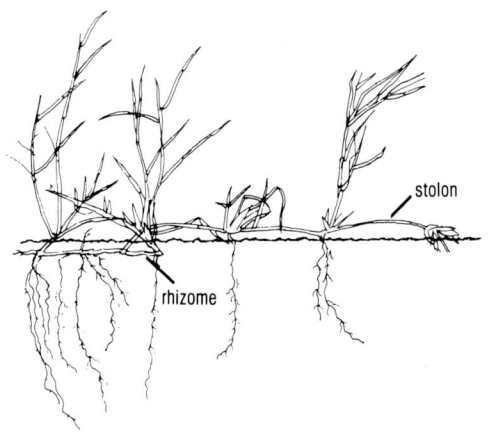

SPREADING STEMS OF PERENNIALS

Figure 51. Terms for vegetative parts of weeds used in identification.

Major Weed Species in Cotton

The photos and descriptions in this book will help you recognize the major weeds in western cotton. Other identification guides listed in the References include the *Growers' Weed Identification Handbook*, which contains photos of more than 200 agricultural weeds. Figure 51 illustrates some terms commonly used in weed descriptions. Learn to recognize seedling weeds as well as mature plants, since it is often too late to make the right management decision by the time weeds are mature. Table 12 shows the approximate germination periods for common annual weeds in the San Joaquin Valley.

Table 12. **Typical Seedling Emergence Dates for Common Annual Weeds in the San Joaquin Valley. The degree to which each weed species competes with cotton is indicated as follows: S = severe competition; M = moderate; L = low; N = nuisance.**

Weed Species	Competitiveness	Months in Which Weed Seedlings Emerge in Cotton Soils
Cheeseweed	N	August through April
Cocklebur	S	May through September
Green Amaranth	S	April through September
Groundcherry	L	June through September
Horseweed	N	all year
Knotweed	N	February through May
Lambsquarters	M	February through May
London rocket	N	August through April
Morningglory	S	May through September
Mustards	N	August through April
Nightshades	S	February through September
Puncturevine	M	April through September
Purslane	M	April through September
Russian thistle	N	February through September
Shepherdspurse	N	August through April
Sowthistle	N	all year
Sunflower	S	February through September
Tumble pigweed	M	March through September
Annual Grasses, most species	M to S	March through September
Volunteer cereals	N	October through June

When using photos and descriptions to identify weeds, remember that most species vary in size and appearance depending on fertility, shade, moisture, and other conditions. Also, it is not possible for a manual to include every species that might grow as a weed in some part of the region. Farm advisors and county agents can usually identify common weeds and they can refer you to specialists for more help.

If you need to submit weeds to a specialist for identification, collect at least three or four examples of each one. Collect the whole plant if possible, including some roots as well as any flowers or fruit that may be present. Collect samples in plastic bags with paper towelling to absorb moisture, and put them in an ice chest to keep them fresh; plant samples deteriorate quickly if left unprotected in the sun or in a vehicle. If you need to keep specimens for more than 1 or 2 days, spread them between layers of newspaper and label each one with the date and place it was collected.

Nutsedges
Cyperus spp.

Nutsedges, also called nutgrasses, are perennial weeds superficially similar to grasses. The most common species in most areas is yellow nutsedge, *Cyperus esculentus*, but purple nutsedge, *C. rotundus*, is a problem in some fields. Nutsedges grow mainly from tubers or "nutlets" formed on rhizomes, mostly in the upper foot of soil. Tubers are commonly spread by farm equipment. Infestations often begin in poorly drained parts of a field.

Nutsedge can reduce yield substantially if allowed to compete with the crop during the first few weeks after planting, especially if the competition is great enough to cause moisture stress in cotton seedlings. To prevent losses, you must stop it from emerging ahead of the crop. Preplant soil-applied herbicides such as dinitroanilines do not provide adequate control, although they may suppress nutsedge growth to some extent, so careful cultivation is essential for early season control.

To give cotton a head start on nutsedge, use a sweep or other shallow cultivating tool to dislodge early nutsedge growth before planting. In preirrigated fields, this cultivation may result in a small loss of soil moisture, but the advantage gained from slowing nutsedge growth is well worth it. After cotton has emerged, use precision equipment to cultivate as closely as possible. Mechanical thinning can also help reduce nutsedge competition.

The foliar-applied herbicides DSMA and MSMA can control nutsedge in the seed row after cotton has emerged. You may need one or two applications, depending on the degree of infestation. It is usually best to wait until cotton plants have two or more true leaves, then use a directed spray aimed at the base of the cotton. However, DSMA can also be applied directly over cotton until the plants are 4 to 6 inches (10 to 15 cm) tall. Both DSMA and MSMA

will cause a purplish discoloration and may retard cotton growth if they contact the growing terminal. Injury to cotton can be severe when plants are stressed for water.

DSMA and MSMA will work as long as they contact the growing point at the base of the nutsedge leaves. You don't need to cover the entire plant. If you treat twice with DSMA/MSMA, the two applications should be 1 to 3 weeks apart, but the latest one must be before first bloom. Don't use these materials more than twice per season; if applied excessively, they may leave a soil residue that can damage some rotation crops.

Metolachlor, a soil-applied herbicide, can provide residual control of yellow nutsedge when used as an early postemergence treatment. It is not effective for purple nutsedge.

An established cotton crop competes effectively with nutsedges, which have little tolerance for shade. If you keep nutsedge in check early in the season, there is no need for special control measures later as long as you have a good stand of cotton. Destroy nutsedge before plants reach the 5- to 6-leaf stage, when they start to produce new tubers.

You can destroy purple nutsedge with repeated summer tillage of dry soil, since the tubers are susceptible to drying. Spring-tooth harrows are usually the best tools for this purpose; discing is often ineffective. Tillage is not likely to be successful in soils that form large clods or in fields where a high water table keeps soil near the surface moist. Tillage is not practical for control of yellow nutsedge, because the tubers can survive up to 4 years in dry soil.

Several combinations of rotation crops and herbicides can make long-term management of nutsedge easier. In heavily infested fields, it may be best to use a rotation crop to reduce nutsedge to a level that can be managed with MSMA/DSMA in cotton. Rotations tested successfully in a recent study in the San Joaquin Valley include: alfalfa treated with EPTC (Eptam); untreated winter barley followed by corn treated with butylate (Sutan+); potatoes treated with EPTC followed by soybeans treated with alachlor (Lasso); potatoes treated with EPTC followed by untreated milo; and 2 years of winter barley followed by summer fallow, with cultivation during the fallow period.

In each case, the 2-year rotation was followed by one season of cotton treated with MSMA/DSMA, and the number of nutsedge tubers and shoots was reduced by at least 95% over the 3-year period. Two hoeings, one 3 weeks after cotton planting and another 3 weeks later, reduced nutsedge by about the same amount as the rotations when this schedule was repeated for 2 years. By comparison, 3 years of continuous cotton treated with MSMA/DSMA reduced nutsedge by 91%. Nutsedge tends to increase in rotations with onions, garlic, peppers, melons, or tomatoes, since the herbicides available in these crops are not effective against it.

A

B

C

Because nutsedge is nearly impossible to eradicate and will quickly reinvade the field if control measures are relaxed, you must maintain a vigorous control program indefinitely.

A. Young nutsedge plants resemble grasses, but the leaves are thicker and stiffer than most grasses. Nutsedge leaves are V-shaped in cross section and are arranged in sets of three at the base, while grass leaves are opposite, in sets of two.

B. Individual nutsedge plants spread to form a dense clump. Flowering stems are triangular in cross section and there are three long, leaflike bracts at the base of each flower head. Mature plants are from 1 to 3 feet (30 to 90 cm) tall.

C. Tubers of yellow nutsedge are produced singly and range in size from ⅛ to ¾ inch (3 to 18 mm) in diameter. They are smooth when mature but have loose scales when smaller. Tubers of purple nutsedge (not shown) are produced in chains, several on a single rhizome. They are oblong and retain their scales when mature. Yellow nutsedge tubers have a pleasant almondlike taste; purple nutsedge tubers are bitter.

Johnsongrass
Sorghum halepense

Originally introduced as a forage crop, johnsongrass is now a major problem in many parts of the Western Region. It reproduces from seed and also spreads from thick rhizomes. An extremely competitive weed, it can reduce cotton yield substantially if not controlled. You can eliminate an infestation over several seasons with careful use of herbicides and tillage, but you must continue sanitation measures indefinitely to prevent reinfestation.

To control established johnsongrass plants, use a combination of tillage and foliar-applied herbicide. Tillage with discs or other implements serves to cut the rhizomes into short segments that are more susecptible to the herbicide. Selective grass herbicides such as fluazifop and sethoxydim are the easiest foliar-applied herbicides to use during the growing season, since they will not injure cotton and do not require special precautions to protect the crop. Apply the first treatment when cotton plants are 4 to 5 inches (10 to 12 cm) tall to reduce early competition. This treatment plus one treatment later will provide adequate control in most cases, but heavily infested fields may need a third application. Make sure the johnsongrass is not drought stressed at the time of treatment.

Glyphosate is effective on johnsongrass, but it can injure cotton if it contacts the foliage. A good way to use glyphosate is to apply it as a preharvest treatment when cotton is mature enough that crop injury will not affect yield. You can also use it to destroy johnsongrass in fencerows and other areas outside the field, and as a spot treatment for small, isolated infestations in the field. You can apply glyphosate with a ropewick or roller applicator to kill johnsongrass that grows taller than the crop, but such applications are not likely to be necessary if selective grass herbicides are properly used. Regardless of the application method, glyphosate works best when johnsongrass is growing vigorously in the heading stage and is not stressed for water. Wet the foliage thoroughly and wait 10 to 14 days before cultivating.

MSMA and DSMA can also kill johnsongrass, but to be effective the entire plant must be treated. Directed sprays, such as those used for nutsedges, are effective only on very small johnsongrass plants that would be covered completely by such applications.

Regrowth from seed is the most common source of reinfestation in fields where established johnsongrass has been destroyed. To prevent seedling growth, apply a dinitroaniline herbicide as a broadcast preplant treatment and incorporate with finishing discs. Use shallow tillage to destroy seedlings that escape the herbicide. Because johnsongrass seeds will continue to germinate in summer after the preplant treatment has dissipated, you will also need a postemergence application of trifluralin to extend control until cotton closes over the rows.

Other measures that help to control johnsongrass include summer fallowing and flooding. Flooding as it is used in rice culture destroys johnsongrass by rotting the roots and rhizomes.

To prevent johnsongrass from reinvading the field, eliminate the weed as thoroughly as possible from roadsides, ditchbanks, and other uncultivated areas around the field. Johnsongrass left uncontrolled in such places serves as a reservoir of seeds that can quickly reverse earlier success in the cultivated area. If possible, install screens to prevent johnsongrass seed from entering the field in irrigation water.

D

E

D. Mature johnsongrass may be 6 feet (2 m) or more tall. Leaf blades are up to 2 feet (60 cm) long, often ½ inch (1 cm) or more wide. The prominent, whitish midvein breaks readily when a leaf is folded. Young plants resemble barnyardgrass, but differ in having a ligule. Johnsongrass flower heads have many branches and are often reddish.

E. The thick, fleshy rhizomes of johnsongrass can produce a new plant from each node.

Bermudagrass
Cynodon dactylon

Bermudagrass is a perennial that spreads from rhizomes and stolons and to a limited extent from seeds. As the stolons spread to new soil, they are at first pressed closely to the surface, but established plants form dense clumps that may be 3 feet (90 cm) tall when supported by other plants such as cotton. Bermudagrass is widespread, but it is most important as a cotton weed in the San Joaquin Valley and the desert valleys.

The selective grass herbicides are effective against bermudagrass. Apply them when the stolons (runners) are about 3 to 9 inches (4 to 14 cm) long. Application methods are essentially the same as for johnsongrass. Glyphosate is useful for spot treatments and for treating bermudagrass outside the planted area. A good time to spot treat is after a final irrigation in late summer. You can treat heavier infestations near harvest, as with johnsongrass. Reinfestation from seed is generally not a problem with bermudagrass.

F.   Bermudagrass spreads as a mat of prostrate stolons and rhizomes. Stolons root at nodes and send up shoots that eventually form a dense clump. The slender spikes of flower heads usually branch from the same point on the stem.

G.   Large crabgrass, *Digitaria sanguinalis*, resembles bermudagrass, but it is an annual that grows from seed and does not spread from stolons or rhizomes. Branches of the flower head arise from two or more points along the stem.

Field Bindweed
Convolvulus arvensis

Field bindweed is widespread in the western U.S., but as a weed in cotton it is most common on loam soils in the western San Joaquin Valley and at higher elevations in Arizona. Often called perennial morningglory, field bindweed spreads from an extensive rootstock as well as from seed. The root system may be 10 feet (3 m) or more deep. The underground stems, or rhizomes, may be several feet long, and they can grow into a series of new plants if chopped up by cultivation when the soil is moist. Seeds remain viable for years.

Management requires a preplant herbicide, tillage, and a midseason application of glyphosate under a hooded sprayer. Glyphosate is most effective on vigorously growing bindweed that has a few flowers, but that has not reached full bloom. When absorbed by vigorous foliage, glyphosate is carried to the roots and rhizomes, destroying part of the rootstock. Multiple treatments are usually necessary to destroy the entire rootstock, so it is difficult to eradicate bindweed from fields planted to cotton every season.

Begin with a broadcast, preplant application of a dinitroaniline. This treatment will provide some control of seedlings and retard the growth of shoots from rootstocks. After cotton has emerged, cultivate to keep new shoots from overtaking the crop. In early June, stop cultivating or skip those spots infested with field bindweed, and let the bindweed grow across the furrows. When the bindweed appears vigorous and has a few flowers, spray it with glyphosate under a hooded sprayer that covers a swath 32 to 34 inches (80 to 85 cm) wide. An irrigation that brings out a flush of new growth will improve control.

Glyphosate will reduce yield if it contacts cotton foliage in mid-season. Attach belting on the bottom of the hood to keep the spray off lower leaves and stems, and make sure the hood is low enough that it will not ride over low branches. A shielded spray is not enough protection.

To destroy bindweed in fallow ground, first disc it, then irrigate to stimulate regrowth. Treat the regrowth with glyphosate when a few flowers are present and disc again in 3 or 4 weeks to destroy any seedlings. You may need to repeat these operations several times to destroy the entire rootstock. Fall treatment with 2,4-D (oil soluble amine formulation) can be effective, although results are more erratic than with glyphosate.

Soil fumigants such as dichloropropene (Telone II), can also provide season-long control, but not eradication, when injected at a depth of 16 to 20 inches (80 to 100 cm). Chisel and disc the soil to eliminate large clods. Fumigate when soil moisture is well below field capacity and when soil temperature is from 60° to 75° F (18° to 24° C). Set the shanks 16 to 20 inches (80 to 100 cm) apart and seal the soil with a roller after injection.

Tillage alone can eliminate field bindweed if repeated often enough in dry soil. Tillage is most effective when shoots are allowed to grow for about 10 days after emergence, since they continue to draw energy from the rootstock before they can produce new food resources to be stored in the rhizomes. Sixteen or more cultivations may be needed over a period of several years. A single deep cultivation of dry soil can set back field bindweed enough that it will not interfere with an annual crop. In fields where deep tillage is needed for another purpose, such as to break up a compacted soil layer, it could take the place of herbicide treatment. Use wide sweeps (reclamation blades) to cut roots and rhizomes about 16 to 18 inches (50 cm) below the surface in dry soil.

Growing alfalfa for 3 or 4 years or rice for 2 years has reduced field bindweed enough in some cases to allow growing an annual crop for one or two seasons. A vigorous alfalfa stand shades out the bindweed, and the flooding in rice culture rots the rootstock. Rotating to a

F

G

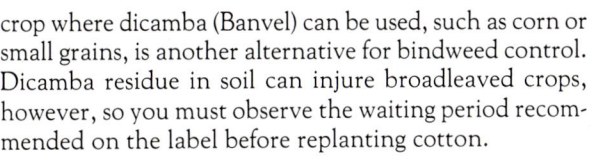

H

I

crop where dicamba (Banvel) can be used, such as corn or small grains, is another alternative for bindweed control. Dicamba residue in soil can injure broadleaved crops, however, so you must observe the waiting period recommended on the label before replanting cotton.

H. Seed leaves of field bindweed are nearly square, with a shallow notch at the tip, while early true leaves are spade shaped. Petioles are flattened and grooved on the upper surface. Plants sprouting from rhizomes lack seed leaves.

I. Trailing stems of field bindweed may be several feet long. Trumpet-shaped white flowers close each afternoon and reopen the following day.

Silverleaf Nightshade
Solanum eleagnifolium

Silverleaf nightshade, also called white horsenettle, is a deep-rooted perennial that spreads from creeping rhizomes as well as by growing from seed. It is widespread in the desert valleys and New Mexico, especially in poorly managed fields. Reduced tillage favors silverleaf nightshade. Once established, it is difficult to eradicate.

Glyphosate is effective on silverleaf nightshade if applied to vigorously growing, mature plants with berries. It is the best spot treatment for small infestations during the growing season. In fields that are generally infested, you can set back silverleaf nightshade with deep tillage, then limit regrowth by planting crops such as alfalfa, corn, sorghum, or small grains that tend to shade it out. Dicamba (Banvel), an herbicide used in corn, sorghum, and small grains, is effective for silverleaf nightshade, but a waiting period is required between application and replanting of cotton or other broadleaved crops susceptible to the soil residue.

J.   Silverleaf nightshade leaves are narrow, gray green, and usually wavy at the edges. Stems usually have slender spines. The blue or purple flowers are about ¾ inch (2 cm) across, with bright yellow stamens; the fruit is a smooth, dull yellow berry. Plants reach a height of about 3 feet (90 cm).

J

Annual Nightshades
Solanum spp.

Nightshades have increased enormously in recent years in the San Joaquin Valley, where they are now a major problem in cotton, beans, tomatoes, melons, and other crops. They also occur in other parts of the Western Region, although infestations elsewhere are generally less severe. Nightshades not only compete with cotton, but their berries can also stain lint if they are present at harvest. Control is complicated by the fact that nightshades germinate and emerge with cotton.

The predominant nightshade species in cotton is black nightshade, *Solanum nigrum*, although hairy nightshade, *S. sarrachoides*, is also abundant in many fields. Each nightshade plant may produce up to 250,000 seeds—so many that a few plants can develop into a dense infestation in just 2 or 3 years. Easily spread on equipment, the sticky seeds are also distributed by birds. Once a nightshade infestation builds up, there may be so many seeds in the soil that a large number of plants survive even when an herbicide treatment is 99% effective. For this reason, it is worthwhile to rogue isolated plants and remove them from the field.

Dinitroanilines will not control nightshades, so you need an extra herbicide such as prometryn. Prometryn works best on nightshades when applied as a preemergence treatment watered in by rain or sprinkler irrigation. Preplant incorporated treatments are also useful, but they are somewhat less dependable than preemergence treatments. Nightshades germinate in the upper 2 inches (5 cm) of soil, so you must keep incorporation shallow for good results. The longer the time between application and rainfall or irrigation, the less effective the treatment is. For adequate crop safety, use prometryn only in combination with a dinitroaniline preplant treatment.

Hoeing may be needed in fields where nightshades survive preplant or preemergence herbicide treatments. Close cultivation and random or synchronous thinning can reduce hoeing costs significantly by limiting weeds to a narrow band along the seed row. Postemergence chemical control of nightshades that come up with cotton is difficult and usually involves some risk of crop injury. Fluometuron with an added surfactant can be applied over the top of small cotton for nightshade control; the treatment delays the crop, but it may be worthwhile as long as good growing conditions after application allow cotton to outgrow the injury rapidly.

Nightshades that germinate after cotton is several inches tall are not difficult to eliminate. Smother seedlings with soil thrown into the seed row in cultivation or use devices such as row weeders to dislodge the seedlings. Directed sprays of cyanazine, diuron, fluometuron, or prometryn can also be effective. For best results, treat when most nightshade seedlings are unfolding their first true leaves. Control is erratic if the weeds are more than about 1 inch (2.5 cm) tall. Nightshades usually are not a problem after cotton closes over the rows as long as the crop is vigorous and the stand is uniform.

The best rotation crop for suppressing nightshade is alfalfa, especially if it is planted in fall. Spring plantings of alfalfa may need a soil-applied herbicide for best nightshade control. Effective herbicides are also available in corn, cereals, beets, lettuce, sorghum, and carrots. Avoid onions, garlic, peppers, and tomatoes.

K. First true leaves of black nightshade are spade shaped with smooth edges and are often purple underneath. Seed leaves are more narrow and pointed.

L. First true leaves of hairy nightshade have wavy edges. Like the older foliage, they usually have many fine, short hairs, especially along the lower side of the main vein.

M. Black nightshade plants are often erect and bushy, like this one, but they may also be sprawling and they vary in color. Hairy nightshade is similar except for the color of the fruit and the fine, dense hairs on foliage and stems.

N. Black nightshade berries turn from green to black when mature and the calyx covers only a small part of the surface. Hairy nightshade berries are green or yellowish when mature and the calyx covers the entire upper surface.

K

L

M

N

O

Control of groundcherries is similar to that described for nightshades, but groundcherries usually emerge later than cotton, so it is easier to control them with directed postemergence sprays. Postemergence treatment with prometryn or diuron is effective for small groundcherry.

O. Groundcherry seed leaves are rounded at the tip and about as wide as long. True leaves are more pointed.

P. Groundcherry flowers are white or yellow, often with green at the center. In some species, petals bend back toward the stem. All groundcherries have a smooth, tomatolike fruit completely enclosed in the calyx. Wright groundcherry, shown here, is usually 1 to 3 feet (30 to 90 cm) tall.

P

Groundcherries
Physalis spp.

Groundcherries occur throughout the Western Region. Wright groundcherry, *Physalis acutifolia* (also known as *P. wrightii*) is the most common species, but lance-leaved groundcherry, *P. lanceifolia*, and tomatillo groundcherry, *P. ixocarpa*, also occur in some areas. Mature plants of all groundcherry species can be recognized by the characteristic fruit.

The closely related small groundcherry, *Chamaesaracha coronopus*, occurs in some New Mexico cotton fields. It closely resembles the *Physalis* species but is smaller and the calyx does not completely enclose the fruit. Unlike the other groundcherries, small groundcherry is a perennial that sprouts from rootstocks as well as from seed.

Annual Morningglories
Ipomoea spp.

Morningglories are a major weed problem in New Mexico and Arizona, and they are also increasing in the San Joaquin Valley. The most common species is Mexican or woolly morningglory, *Ipomoea hirsutula*. Others are tall morningglory, *I. purpurea*, scarlet morningglory, *I. coccinea*, and ivyleaf morningglory, *I. hederacea*. Some species are cultivated for their showy flowers, but they become troublesome weeds when they escape from gardens.

Annual morningglories are twining plants with stems several feet long. They climb over cotton plants, forming dense mats that interfere with defoliation and harvest. Seeds germinate at a depth of 4 inches (10 cm) or more, much deeper than most annuals. Most seedlings emerge following irrigations, but they may also appear between cotton emergence and first irrigation, when surface soil is too dry to allow germination of other annuals. Control of annual morningglories is critical from crop emergence to the second or third irriation, when the weeds are most competitive. Destroy the seedlings while they are small, because once they have twined up cotton stems they are difficult to control without injuring the crop.

For early season control of annual morningglories, use a combination of a dinitroaniline with prometryn as a preplant incorporated treatment, or apply prometryn preemergence where a dinitroaniline was incorporated earlier. Dinitroanilines alone are not effective. Postemergence treatments may also be necessary. Directed sprays of fluometuron, cyanazine, diuron, or prometryn are effective. MSMA applied as a directed spray with a wetting agent is effective if applied when seedlings are small enough for the spray to cover the whole plant. MSMA works better in combination with diuron or prometryn.

Close cultivation can destroy or bury most seedlings if performed while they are small. Bezzerides row weeders and Texas rod weeders are very effective once cotton is well established. However, some hoeing is usually needed to assure complete control of annual morningglories, even if herbicides and cultivation have been used effectively.

Q. Seedlings of annual morningglories are similar to those of field bindweed, but seed leaves have a much deeper notch.

R. Annual morningglory flowers vary from violet or blue to pink and red. Some are mostly white with only small markings of other colors. All are similar in shape.

Other Broadleaved Weeds

Weeds of the amaranth family are among the most abundant summer annuals in most fields. The most common species in the San Joaquin Valley are redroot pigweed, *Amaranthus retroflexus*, green amaranth, *A. hybridus*, and tumble pigweed, *A. albus*. Palmer amaranth, *A. palmeri*, is the dominant species in Arizona. These weeds are all susceptible to dinitroaniline herbicides, but they sometimes germinate in summer, after the preplant treatment has dissipated or when treated soil has been removed from furrows. Green amaranth and redroot pigweed may quickly grow so large that they overtop cotton. Lambsquarters, another abundant annual, is often confused with pigweeds but can be distinguished by the scaly coating on the young foliage. It germinates earlier than most pigweeds.

Summer annuals more difficult to control include sunflower, *Helianthus annuus*, and cocklebur, *Xanthium strumarium*. Dinitroanilines do not control these species, so you must add a more effective material, such as prometryn, to your herbicide program where these weeds are abundant.

Jimsonweed, *Datura stramonium*, is a nightshade family annual that is increasing in some areas. *D. meteloides*, known as tolguacha, is a perennial that is becoming more common in San Joaquin Valley cotton. Both have large, gray green leaves that have a strong odor when crushed. Large, tubular white flowers and spiny seed pods make these weeds easy to recognize when mature; both species are illustrated in the *Growers' Weed Identification Handbook*. Foliar sprays of DSMA or MSMA are effective for jimsonweed.

Spurred anoda, *Anoda cristata*, is a weed in the same family as cotton that is increasing in New Mexico and is potentially troublesome in other areas also. Prometryn, cyanazine, and fluometuron are effective controls.

Winter annuals are generally less important as weeds in cotton, since they are usually destroyed by preplant tillage. However, several species resistant to dinitroanilines can pose a problem where dinitroanilines are applied as a preplant treatment in fall or winter. Resistant

Q

R

winter weeds that emerge after the treatment, including volunteer cereals, can deplete soil moisture and may interfere with seedbed preparation. In fields where beds are prepared in fall for spring planting, an appropriate herbicide may be needed to destroy winter weeds, especially if rainfall makes it impossible to cultivate before planting. Aerial application is often preferable, since it avoids soil compaction.

Mustard family annuals such as wild mustards, *Brassica* spp.; shepherdspurse, *Capsella bursa-pastoris*; and London rocket, *Sisymbrium irio*, are common winter weeds

S

T

U

in most areas and are resistant to dinitroanilines. Other resistant winter weeds include prickly lettuce, *Lactuca serriola*; sowthistles, *Sonchus* spp.; and cheeseweed, *Malva parviflora*. These species are all illustrated in the *Growers' Weed Identification Handbook.*

S. Seedlings of all common pigweeds are similar. Seed leaves are long and narrow and are often red underneath. First true leaves have a shallow notch at the tip that distinguishes them from seedlings of nightshades and groundcherries.

T. Leaves of redroot pigweed (left) are large and rough to the touch. Leaf blades are often 3 or 4 inches (7 to 10 cm) long and petioles may be nearly as long as the blade. Flowers are mostly in dense spikes at the tops of the main stem and branches. Stems are deeply grooved. Plants are often several feet tall. Leaves of tumble pigweed (right) seldom exceed 2 inches (5 cm) in length. Stems are smooth and white. Flowers are in small clusters in leaf axils; there are no dense spikes at the tops of stems. Plants are usually no more than 4 feet (1.2 m) tall.

U. Green amaranth, found in the San Joaquin Valley, and Palmer amaranth, found in the desert valleys, have long, slender, leafless flower spikes. Plants may be up to 8 feet (2.4 m) tall.

V. Seed leaves of lambsquarters are straplike, with edges nearly parallel. Seed leaves and early true leaves are often purple underneath. Lambsquarters foliage, especially younger leaves, is coated with tiny white scales that look like flour to the naked eye. The scaly leaf texture distinguishes lambsquarters from most other seedlings, including nightshades, groundcherries, and pigweeds.

W. Small, greenish flowers of lambsquarters are packed in clusters at the tips of the main stem and branches. Plants may be from a few inches to several feet tall when mature.

X. Seed leaves of cocklebur are narrow, pointed, and shiny green. True leaves are shallowly toothed along the edges.

Y. Mature cocklebur plants are usually 2 to 5 feet (0.6 to 1.5 m) tall. Leaves are rough to the touch and petioles are much longer than blades. Pollen-producing male flowers are in globular heads at the tips of stems. Female flowers, located in leaf axils, form a bur covered with hooked spines. Burs are usually about an inch long.

Z. Mature spurred anoda can be as tall as cotton. Upper leaves are lance shaped; lower leaves are broader, though pointed. Leaves are hairy, especially on lower surface, and there is often a purple mark on the main vein.

V

X

W

Y

Z

A

A. London rocket may reach a height of up to 4 feet (1.2 m). Flower heads are at the tips of stems, and the slender seed pods are 1½ to 2½ inches (4 to 7 cm) long. Leaves are pointed, spade shaped, and divided into two or more sets of lobes at the base. Check London rocket for lygus bugs early in the season.

B. The distinctive, triangular seed pods of shepherdspurse are on short stalks radiating from the flowering stem. Plants are up to about 20 inches (50 cm) tall. Shepherdspurse is an important host of lygus bugs and false chinch bugs.

C. Wild mustards, such as *Brassica kaber* shown here, are common winter weeds and important hosts of lygus bugs. Depending on the species, mustards may be several feet tall. Leaves, especially those near the base of the plant, are often deeply lobed.

Annual Grasses

Dinitroaniline herbicides generally provide good control of most annual grasses, although barnyardgrass, *Echinochloa crus-galli*; junglerice, *E. colonum*; and large crabgrass, *Digitaria sanguinalis*, germinate throughout the season and may appear after a preplant treatment has dissipated. You can control late-emerging grasses with cultivation and selective grass herbicides.

B

C

D. Stems of young barnyardgrass plants often grow outward along the ground before turning upward, and commonly root at inner nodes. Stems are flattened near the base.

E. To recognize barnyardgrass, strip off leaves and look at the collar region. Aside from junglerice, barnyardgrass is the only common summer annual grass that lacks a ligule.

F. Mature barnyardgrass can be up to 6 feet (2 m) tall in moist situations, but may be only 6 inches (15 cm) tall in dry places. Flower heads vary, but they are usually drooping and lower branches are further apart than upper ones.

G. Junglerice is similar to barnyardgrass, but usually has reddish stripes across the leaf blades.

D

E

F

G

Appendix: Degree-Day Tables

With a record of daily high and low temperatures, you can obtain degree-day values from the tables presented here. These tables correspond to the four models of cotton development commonly used in the Western Region. Each one is based on a different developmental threshold and/or upper temperature limit, so it is important to use the same table each time.

The values in Tables 14, 15, and 16 are for 24-hour periods; to obtain a degree-day value for a given day, simply find the figure in the table that corresponds to the high and low temperatures for that day. For odd-numbered temperatures not shown in the table, use the average of the degree-day values for the two closest even-numbered temperatures.

To use Table 13, you must divide each day into 12-hour segments, find the appropriate value for each segment, then add them to obtain the total for the day. For example, if you need a degree-day total for 3 consecutive

MINIMUM TEMPERATURES

MAX TEMPS	46	48	50	52	54	56	58	60	62	64	66	68	70	72	74	76	78	80	82	84	86	88	90
60	0	0	0	0	0	0	0	0															
62	0	0	0	0	0	0	0	1	1														
64	0	0	0	0	0	1	1	1	2	2													
66	0	1	1	1	1	1	1	2	2	3	3												
68	1	1	1	1	1	1	2	2	3	3	4	4											
70	1	1	1	1	2	2	2	3	3	4	4	5	5										
72	1	2	2	2	2	2	3	3	4	4	5	5	6	6									
74	2	2	2	2	2	3	3	4	4	5	5	6	6	7	7								
76	2	2	2	3	3	3	4	4	5	5	6	6	7	7	8	8							
78	3	3	3	3	3	4	4	5	5	6	6	7	7	8	8	9	9						
80	3	3	3	4	4	4	5	5	6	6	7	7	8	8	9	9	10	10					
82	3	4	4	4	4	5	5	6	6	7	7	8	8	9	9	10	10	11	11				
84	4	4	4	5	5	5	6	6	7	7	8	8	9	9	10	10	11	11	12	12			
86	4	4	5	5	5	6	6	7	7	8	8	9	9	10	10	11	11	12	12	13	13		
88	5	5	5	5	6	6	7	7	8	8	9	9	10	10	11	11	12	12	13	13	14	14	
90	5	5	6	6	6	7	7	8	8	9	9	10	10	11	11	12	12	13	13	14	14	15	15
92	6	6	6	6	7	7	8	8	9	9	10	10	11	11	12	12	13	13	14	14	15	15	16
94	6	6	7	7	7	8	8	9	9	10	10	11	11	12	12	13	13	14	14	15	15	16	16
96	6	7	7	7	8	8	9	9	10	10	11	11	12	12	13	13	14	14	15	15	16	16	17
98	7	7	8	8	8	9	9	10	10	11	11	12	12	13	13	14	14	15	15	16	16	17	17
100	7	8	8	8	9	9	10	10	11	11	12	12	13	13	14	14	15	15	16	16	17	17	18
102	8	8	8	9	9	10	10	11	11	12	12	13	13	14	14	15	15	16	16	17	17	18	18
104	8	9	9	9	10	10	11	11	12	12	13	13	14	14	15	15	16	16	17	17	18	18	19
106	9	9	9	10	10	11	11	12	12	13	13	14	14	15	15	16	16	17	17	18	18	19	19
108	9	10	10	10	11	11	12	12	13	13	14	14	15	15	16	16	17	17	18	18	19	19	20
110	10	10	10	11	11	12	12	13	13	14	14	15	15	16	16	17	17	18	18	19	19	20	20
112	10	11	11	11	12	12	13	13	14	14	15	15	16	16	17	17	18	18	19	19	20	20	21
114	11	11	11	12	12	13	13	14	14	15	15	16	16	17	17	18	18	19	19	20	20	21	21
116	11	12	12	12	13	13	14	14	15	15	16	16	17	17	18	18	19	19	20	20	21	21	22
118	12	12	12	13	13	14	14	15	15	16	16	17	17	18	18	19	19	20	20	21	21	22	22

Table 13. Degree-Day Values for 12-Hour Periods, Based on Developmental Threshold of 60° F with No Maximum Temperature.

days with the high and low temperatures shown below, start by finding the value in Table 13 corresponding to a low of 48° F and a high of 80° F, the first 12-hour period in day 1. The next 12-hour period is from the high of day 1 to the low of day 2; the value for 52° F and 80° F is 4 °D. To obtain a total for the entire period, continue these steps, adding the results each time. Remember to include the period between the high on day 3 and the low on day 4 to get a total for the full 3 days.

		Degree-day values from Table 13	Cumulative total
DAY 1			
low	48° F		
high	80	3	3
DAY 2			
low	52	4	7
high	86	5	12
DAY 3			
low	57	6	18
high	84	5.5	23.5
DAY 4			
low	59	6	29.5
high	90		

MINIMUM TEMPERATURES

MAX TEMPS	46	48	50	52	54	56	58	60	62	64	66	68	70	72	74	76	78	80	82	84	86	88	90
60	1	1	1	2	2	2	3	4	5														
62	1	2	2	2	3	3	4	5	6	7													
64	1	3	3	3	4	4	5	6	7	8	9												
66	1	4	4	4	5	5	6	7	8	9	10	11											
68	1	5	5	5	6	6	7	8	9	10	11	12	13										
70	1	5	6	6	7	7	8	9	10	11	12	13	14	15									
72	1	6	7	7	8	8	9	10	11	12	13	14	15	16	17								
74	1	7	8	8	8	9	10	11	12	13	14	15	16	17	18	19							
76	1	8	9	9	9	10	11	12	13	14	15	16	17	18	19	20	21						
78	1	9	9	10	10	11	12	13	14	15	16	17	18	19	20	21	22	23					
80	1	10	10	11	11	12	13	14	15	16	17	18	19	20	21	22	23	24	25				
82	1	11	11	12	12	13	14	15	16	17	18	19	20	21	22	23	24	25	26	27			
84	1	12	12	13	13	14	15	16	17	18	19	20	21	22	23	24	25	26	27	28	29		
86	1	13	13	14	14	15	16	17	18	19	20	21	22	23	24	25	26	27	28	29	30	31	
88	1	24	25	25	26	27	17	18	19	20	21	22	23	24	25	26	27	28	29	29	30	31	31
90	1	23	24	24	25	26	17	18	19	20	21	22	23	24	25	26	27	28	29	30	30	31	31	31
92	1	23	23	24	24	25	18	19	20	21	22	23	24	25	26	26	27	28	29	30	31	31	31	31
94	1	22	23	23	24	25	18	19	20	21	22	23	24	25	26	27	28	28	29	30	31	31	31	31
96	1	22	22	23	23	24	19	20	21	22	23	23	24	25	26	27	28	29	29	30	31	31	31	31
98	1	22	22	23	23	24	19	20	21	22	23	24	25	25	26	27	28	29	29	30	31	31	31	31
100	1	22	22	22	23	23	20	20	21	22	23	24	25	26	27	27	28	29	30	30	31	31	31	31
102	1	21	22	22	22	23	20	21	22	23	23	24	25	26	27	27	28	29	30	30	31	31	31	31
104	1	21	21	22	22	23	20	21	22	23	24	24	25	26	27	28	28	29	30	30	31	31	31	31
106	1	21	21	22	22	22	20	21	22	23	24	25	25	26	27	28	28	29	30	30	31	31	31	31
108	1	21	21	21	22	22	21	21	22	23	24	25	26	26	27	28	29	29	30	30	31	31	31	31
110	1	21	21	21	21	22	21	22	23	23	24	25	26	27	27	28	29	29	30	30	31	31	31	31
112	1	20	21	21	21	22	21	22	23	24	24	25	26	27	27	28	29	29	30	30	31	31	31	31
114	1	20	21	21	21	21	21	22	23	24	24	25	26	27	27	28	29	29	30	30	31	31	31	31
116	1	20	20	21	21	21	21	22	23	24	25	25	26	27	28	28	29	29	30	30	31	31	31	31
118	1	20	20	20	21	21	22	22	23	24	25	26	26	27	28	28	29	29	30	30	31	31	31	31

Table 14. Degree-Day Values for 24-Hour Periods, Based on Developmental Threshold of 55° F and Maximum Temperature of 86° F.

MINIMUM TEMPERATURES

MAX TEMPS	46	48	50	52	54	56	58	60	62	64	66	68	70	72	74	76	78	80	82	84	86	88	90
60 1	2	2	2	3	4	5	6	7															
62 1	3	3	3	4	5	6	7	8	9														
64 1	4	4	4	5	6	7	8	9	10	11													
66 1	5	5	5	6	7	8	9	10	11	12	13												
68 1	5	6	6	7	8	9	10	11	12	13	14	15											
70 1	6	7	7	8	9	10	11	12	13	14	15	16	17										
72 1	7	8	8	9	10	11	12	13	14	15	16	17	18	19									
74 1	8	9	9	10	11	12	13	14	15	16	17	18	19	20	21								
76 1	9	10	10	11	12	13	14	15	16	17	18	19	20	21	22	23							
78 1	10	11	11	12	13	14	15	16	17	18	19	20	21	22	23	24	25						
80 1	11	12	12	13	14	15	16	17	18	19	20	21	22	23	24	25	26	27					
82 1	12	13	13	14	15	16	17	18	19	20	21	22	23	24	25	26	27	28	29				
84 1	13	14	14	15	16	17	18	19	20	21	22	23	24	25	26	27	28	29	30	31			
86 1	14	14	15	16	17	18	19	20	21	22	23	24	25	26	27	28	29	30	31	32	33		
88 1	15	15	16	17	18	19	20	21	22	23	24	25	26	27	28	29	30	31	32	33	34	35	
90 1	16	16	17	18	19	20	21	22	23	24	25	26	27	28	29	30	31	32	33	34	35	36	37
92 1	17	17	18	19	20	21	22	23	24	25	26	27	28	29	30	31	32	33	34	35	36	37	38
94 1	18	18	19	20	21	22	23	24	25	26	27	28	29	30	31	32	33	34	35	36	37	38	39
96 1	19	19	20	21	21	22	23	24	25	26	27	28	29	30	31	32	33	34	35	36	37	38	39
98 1	19	20	20	21	22	23	24	25	26	27	28	29	30	31	32	33	34	35	36	37	38	38	39
100 1	20	20	21	22	23	24	25	26	27	28	29	29	30	31	32	33	34	35	36	37	38	39	39
102 1	20	21	22	22	23	24	25	26	27	28	29	30	31	32	33	34	35	35	36	37	38	39	40
104 1	21	21	22	23	24	25	26	27	27	28	29	30	31	32	33	34	35	36	37	37	38	39	40
106 1	21	22	23	23	24	25	26	27	28	29	30	31	32	32	33	34	35	36	37	38	38	39	40
108 1	22	22	23	24	24	25	26	27	28	29	30	31	32	33	34	34	35	36	37	38	38	39	40
110 1	22	23	23	24	25	26	27	28	29	29	30	31	32	33	34	35	36	36	37	38	39	39	40
112 1	23	23	24	24	25	26	27	28	29	30	31	31	32	33	34	35	36	36	37	38	39	39	40
114 1	23	23	24	25	26	26	27	28	29	30	31	32	33	33	34	35	36	37	37	38	39	39	40
116 1	23	24	24	25	26	27	28	28	29	30	31	32	33	34	34	35	36	37	37	38	39	39	40
118 1	24	24	25	25	26	27	28	29	30	30	31	32	33	34	35	35	36	37	38	38	39	39	40

Table 15. Degree-Day Values for 24-Hour Periods, Based on
Developmental Threshold of 53.4° F and Maximum
Temperature of 94° F.

MINIMUM TEMPERATURES

MAX TEMPS	46	48	50	52	54	56	58	60	62	64	66	68	70	72	74	76	78	80	82	84	86	88	90
60 1	1	1	2	2	2	3	4	5															
62 1	2	2	2	3	3	4	5	6	7														
64 1	3	3	3	4	4	5	6	7	8	9													
66 1	4	4	4	5	5	6	7	8	9	10	11												
68 1	5	5	5	6	6	7	8	9	10	11	12	13											
70 1	5	6	6	7	7	8	9	10	11	12	13	14	15										
72 1	6	7	7	8	8	9	10	11	12	13	14	15	16	17									
74 1	7	8	8	8	9	10	11	12	13	14	15	16	17	18	19								
76 1	8	9	9	9	10	11	12	13	14	15	16	17	18	19	20	21							
78 1	9	9	10	10	11	12	13	14	15	16	17	18	19	20	21	22	23						
80 1	10	10	11	11	12	13	14	15	16	17	18	19	20	21	22	23	24	25					
82 1	11	11	12	12	13	14	15	16	17	18	19	20	21	22	23	24	25	26	27				
84 1	12	12	13	13	14	15	16	17	18	19	20	21	22	23	24	25	26	27	28	29			
86 1	13	13	14	14	15	16	17	18	19	20	21	22	23	24	25	26	27	28	29	30	31		
88 1	14	14	15	15	16	17	18	19	20	21	22	23	24	25	26	27	28	29	30	31	32	33	
90 1	15	15	16	16	17	18	19	20	21	22	23	24	25	26	27	28	29	30	31	32	33	34	35
92 1	16	16	17	17	18	19	20	21	22	23	24	25	26	27	28	29	30	31	32	33	34	35	36
94 1	17	17	18	18	19	20	21	22	23	24	25	26	27	28	29	30	31	32	33	34	35	36	37
96 1	18	18	19	19	20	21	22	23	24	25	26	27	28	29	30	31	32	33	34	35	36	37	38
98 1	19	19	20	20	21	22	23	24	25	26	27	28	29	30	31	32	33	34	35	36	37	38	39
100 1	20	20	21	21	22	23	24	25	26	27	28	29	30	31	32	33	34	35	36	37	38	39	40
102 1	21	21	22	22	23	24	25	26	27	28	29	30	31	32	33	34	35	36	37	38	39	40	41
104 1	22	22	23	23	24	25	26	27	28	29	30	31	32	33	34	35	36	37	38	39	40	41	42
106 1	23	23	24	24	25	26	27	28	29	30	31	32	33	34	35	36	37	38	39	40	41	42	43
108 1	23	24	25	25	26	27	28	29	30	31	32	33	34	35	36	37	38	39	40	41	42	43	44
110 1	24	25	26	26	27	28	29	30	31	32	33	34	35	36	37	38	39	40	41	42	43	44	45
112 1	25	26	27	27	28	29	30	31	32	33	34	35	36	37	38	39	40	41	42	43	44	45	46
114 1	26	27	28	28	29	30	31	32	33	34	35	36	37	38	39	40	41	42	43	44	45	46	47
116 1	27	28	29	29	30	31	32	33	34	35	36	37	38	39	40	41	42	43	44	45	46	47	48
118 1	28	29	30	30	31	32	33	34	35	36	37	38	39	40	41	42	43	44	45	46	47	48	49

Table 16. Degree-Day Values for 24-Hour Periods, Based on Developmental Threshold of 55° F with No Maximum Temperature.

References

Cotton Growth and Development

Growth and Development of the Cotton Plant in Arizona. Publication No. 8168 (Arizona**).

Morphology of the Cotton Plant, by J. R. Mauney, in Advances in Production and Utilization of Quality Cotton: Principles and Practices; F. C. Elliot, M. Hoover, and W. K. Porter, Jr., editors, 1968, 532 pp., Iowa State University Press, Ames.

A Slide Rule for Cotton Crop and Insect Management. Leaflet 21361 (California*).

Soil, Water, and Nutrients

Safe Handling and Use of Anhydrous Ammonia. WRAES 37.†

Safe Handling and Use of Aqua Ammonia. WRAES 45.†

Soil and Plant Tissue Testing in California. Bulletin 1879 (California*).

Water Management for Cotton. Bulletin 1904 (California*).

Western Fertilizer Handbook, 6th edition, 1980, 269 pp. Soil Improvement Committee, California Fertilizer Association. Order from Interstate Printers and Publishers, Inc., Danville, IL 61832.

Degree-Days and Weather

Average Daily Air Temperatures and Precipitation in California. Special Publication 3285 (California*).

Construction of Weather Shelters. Publication 2371 (California*).

Degree-Days: The Calculation and Use of Heat Units in Pest Management. Publication 21373 (California*).

The Grower's Weather Guide for Farming Practices, by J. Y. Wang and others. 1982, 66 pp. Milieu Information Services, Inc., P.O. Box 6536, San Jose, CA 95150.

Cultural Practices

Effects of Row Spacing on Cotton Yield, Quality, and Plant Characteristics. Publication 1903 (California*).

Harvesting Cotton with Spindle-type Pickers. Publication 2814. (California*).

Plant Growth Regulators; a Study Guide for Agricultural Pest Control Advisers. Publication 4047 (California*).

Study Guide for Agricultural Pest Control Advisers on Defoliation and Other Harvest-aid Practices. Publication 4043 (California*).

Pesticide Application and Safety

Calibration of Pesticide Application Equipment. Publication Q362 (Arizona**).

How Much Chemical Do You Put in the Tank? Leaflet 2718 (California*).

Pesticide Application and Safety Training. Publication 4070 (California*).

Pesticide Safety Information Series. California Department of Food and Agriculture, 1220 'N' Street, Sacramento, CA 95814. Includes specific procedures for handling hazardous pesticides.

Reducing Pesticide Hazards to Honey Bees: Mortality Prediction Techniques and Integrated Management Strategies. Leaflet 2883 (California*).

Study Guide for Agricultural Pest Control Advisers on the Safe Application of Agricultural Chemicals—Equipment and Calibration. Publication 4048 (California*).

Vertebrates

Study Guide for Agricultural Pest Control Advisers on Vertebrate Pests. Publication 4049 (California*).

Vertebrate Pest Control Handbook. California Department of Food and Agriculture, 1220 'N' Street, Sacramento, CA 95814.

Insects

Arizona Cotton Insects. Bulletin A23 R (Arizona**).

Cotton Insect Management Suggestions. Publication 400 J-7 (New Mexico***).

Identification of Lepidopterous Larvae Attacking Cotton, with illustrated key (primarily California species), by G. T. Okumura. California Department of Agriculture, Special Publication No. 282, 1962.

Insect and Nematode Recommendations for Cotton. Leaflet 2083 (California*).

Insect Pest Management for Cotton. (unnumbered) (Arizona**).

Pest Management Guide for Insects and Nematodes of Cotton in California. Publication 4089 (California*).

Pink Bollworm Control in the Western United States. U.S. Department of Agriculture, Science and Education Administration, ARM-W-16, 1980.

A Slide Rule for Cotton Crop and Insect Management. Leaflet 21361 (California*).

Mites and Mite Sampling

Presence-Absence Sampling of Spider Mite Densities on Cotton. California Agriculture, July-August 1981, p. 10.

Spider Mite Pests of Cotton. Leaflet 2888 (California*).

Within-Plant Distribution of Spider Mites on Cotton: A Developing Implementable Monitoring Program. Environmental Entomology, vol. 12, No. 1, pp. 128–134, 1983.

Nematodes and Diseases

Aflatoxins in Cottonseed. Publication Q422 (Arizona**).

Compendium of Cotton Diseases. American Phytopathological Society, 1981, 87 pp. 3340 Pilot Knob Road, St. Paul, MN 55121.

Cotton Seedling Diseases. Publication Q409 (Arizona**).

Growing Cotton on Verticillium-Infested Land. Cooperative Extension Service Circular 445 (New Mexico***).

Leaf Diseases of Cotton. Publication Q411 (Arizona**).

Plant Nematology, An Agricultural Training Aid. S. M. Ayoub, California Department of Food and Agriculture, 1977.

Plant Pathology, by G. N. Agrios. 2nd edition, 1978, 703 pp. Academic Press, New York and San Francisco.

Root-Knot Nematode on Cotton. Leaflet 2819 (California*).

Soil Solarization: A Nonchemical Method for Controlling Diseases and Pests. Publication 21377 (California*).

Study Guide for Agricultural Pest Control Advisers on Nematodes and Nematicides. Publication 4045 (California*).

Study Guide for Agricultural Pest Control Advisers on Plant Diseases. Publication 4046 (California*).

Texas Root Rot of Cotton. Publication Q410 (Arizona**).

Verticillium Wilt of Cotton. Publication Q408 (Arizona**).

---

*California publications available from Agriculture and Natural Resources Publications, University of California, 6701 San Pablo Avenue, Oakland, CA 94608. Free catalog on request.

**Arizona publications available from University of Arizona College of Agriculture, Cooperative Extension Service Publications Section, Tucson, AZ 85721.

***New Mexico publications available from New Mexico State University Cooperative Extension Service, Las Cruces, NM 88003.

†Western Regional Agricultural Extension Service publications are available in each state at the above addresses.

pest resurgence. an increase in a pest population following a pesticide application intended to reduce it (page 31).

pesticide resistance. the genetically acquired ability of an organism to survive a pesticide application at doses that once killed most individuals of the same species (Figure 18).

petiole. a stalk connecting a leaf to a stem.

pheromone. a substance secreted by an organism to affect the behavior or development of other organisms of the same species; in this book, a chemical released by an insect to attract others of the same species. Also, a synthetic chemical with the same structure and activity as a natural pheromone.

phloem. tissue that conducts the products of photosynthesis through a plant (Figure 3).

photosynthesis. the process by which plants use the energy of sunlight to build carbon dioxide into sugars and other carbohydrates.

phytotoxicity. injury to plants caused by exposure to a chemical.

Pima cotton. *Gossypium barbadense.*

pinhead square. a square approximately ⅛ inch (3mm) or less in length.

postemergence herbicide. in this book, an herbicide applied after crop emergence; also, an herbicide applied after emergence of target weeds.

preemergence herbicide. in this book, an herbicide applied between planting and crop emergence; also, an herbicide applied before emergence of weeds.

proleg. a fleshy, unsegmented leg of caterpillars (Figure 32).

pupa. a nonfeeding, inactive stage in which the tissues of an insect larva are formed into those of the adult.

pupate. to molt from the larval stage to the pupa.

quadrant. one of four equal parts into which a field is divided for monitoring.

reservoir. a stock of inoculum or a population of organisms which can survive outside the field or in the absence of a host, but which may enter a field or contact a host under certain conditions.

resistant. able to tolerate conditions harmful to other strains of the same species. In weed science, this term is applied to weeds not susceptible to particular herbicides. See also pesticide resistance.

resurgence. see pest resurgence.

rhizome. a horizontal, underground stem, especially one that roots at the nodes to produce new plants (Figure 51).

rootstock. the underground portion of a plant, such as roots and rhizomes, capable of producing new plants when the aboveground portion is destroyed.

rosetted bloom. a flower whose petals have been tied together with silk by a pink bollworm larva.

sclerotium (plural, **sclerotia**). a compact mass of hardened mycelium that serves as a dormant stage in some fungi.

secondary infection. infection by a pathogen that enters the host through an injury caused by another pathogen.

secondary pest outbreak. see p. 31.

seedcotton. harvested lint that is still attached to seeds; i.e., the lint before ginning.

seed leaf. cotyledon.

senescence. the stage in the life cycle of a plant or plant part from maturity to death, in which metabolic products accumulate, respiratory rate increases, and dry weight decreases.

sequential sampling. a sampling method in which the number of samples is not fixed in advance (page 35).

side-dressing. fertilizer or other material added to the soil around a growing crop.

skeletonize. to remove leaf tissue between the veins, leaving the network of veins intact.

spiracle. an external opening of the system of ducts, or tracheae, that serves as a respiratory system in insects.

spore. a reproductive body usually consisting of one or a few cells, produced by fungi and certain other organisms.

square. a cotton flower bud.

stolon. a stem that grows horizontally along the ground, producing roots and shoots at the nodes (Figure 51).

stoma (plural, **stomata or stomates**). an opening in the surface of a leaf or other plant parts that permits the passage of gases and water vapor.

stub cotton. a cotton crop in which the stalks are cut down after harvest, but the crown and rootstock are left in the ground to regrow the following season.

substituted dinitroanilines. a class of herbicides widely used for preplant application in cotton. (page 116).

systemic pesticide. a pesticide that moves through a plant from the point of application, usually via the vascular system.

tail water. irrigation water that has drained from a field.

target pest. a pest species that a control action is intended to destroy.

tensiometer. a device for measuring soil moisture, consisting of a buried tube of water that develops a partial vacuum as surrounding soil dries out.

terminal. the growing tip of a stem, especially the main stem.

Texas root rot. Phymatotrichum root rot (page 104).

thorax. the second of three major divisions in the body of an insect, and the one bearing the legs and wings.

threshold. see developmental threshold or treatment threshold.

thurberia weevil. a race of the boll weevil that feeds on wild cotton (page 59).

tolerance. inherent lack of susceptibility to a pesticide. Also, the ability of a plant to grow in spite of infection by a pathogen.

top crop. fruit produced in the second fruiting cycle of cotton (page 15), mainly on upper branches.

toxin. a poisonous substance produced by a living organism and composed of protein.

transpiration. the evaporation of water vapor from plants, mostly through stomata.

trap crop. a crop or portion of a crop intended to attract pests so they can be destroyed by treating a relatively small area or by destroying the trap crop and the pests together.

treatment threshold. the level of pest population at which a pesticide or other control measure is needed to prevent eventual economic injury to the crop.

true leaf. any leaf produced after the cotyledons.

upland cotton. *Gossypium hirsutum.*

vascular system. the system of plant tissues that conducts water, mineral nutrients, and products of photosynthesis through the plant, consisting of the phloem and xylem.

vector. an organism able to transport and transmit a pathogen to a host.

vegetative growth. the growth of roots, stems, and leaves, as distinguished from development of flowers and fruit.

xylem. tissue that conducts water and nutrients from the roots up through the plant (Figure 3).

A

B

C

D

E

F

G

H

I

4